Practical Schedule Risk Analysis

Practical Schedule Risk Analysis

DAVID T. HULETT, PhD

Routledge
Taylor & Francis Group

LONDON AND NEW YORK

First published in paperback 2024

First published 2009 by Gower Publishing

First published 2016
by Routledge
4 Park Square, Milton Park, Abingdon, Oxon OX14 4RN

and by Routledge
605 Third Avenue, New York, NY 10158

Routledge is an imprint of the Taylor & Francis Group, an informa business

British Library Cataloguing in Publication Data
Hulett, David

    Practical schedule risk analysis
    1. Production scheduling 2. Production scheduling -
    Mathematical models 3. Risk assessment
    I. Title
    658.5'3

Library of Congress Cataloging-in-Publication Data
Hulett, David.
    Practical schedule risk analysis / by David Hulett.
      p. cm.
    Includes bibliographical references and index.
    ISBN 978-0-566-08790-5
    1. Corporate culture. 2. Risk assessment. I. Title.
    HD58.7.H838 2009
    658.15'5--dc22

                                                   2008049792

ISBN: 978-0-566-08790-5 (hbk)
ISBN: 978-1-03-283796-3 (pbk)
ISBN: 978-1-315-60188-5 (ebk)

DOI: 10.4324/9781315601885

Contents

List of Figures

List of Tables

Preface

One secret of success is to know people who know more than you do. I've been working in risk management for over twenty-five years, and I've learned a lot in that time. I've developed my own specialist areas of expertise, including management of positive upside risk or opportunity, understanding the influence of risk attitude on decision-making, and the role of risk in society. But when it comes to quantitative risk analysis I know where to turn. Dr David Hulett is internationally recognized as a world authority on the subject, and there is no-one better than him to explain its intricacies and subtleties. I'm pleased to count David as a friend and colleague and we've worked closely together over the years, so it's easy for me to tap into his wealth of knowledge and practical expertise whenever I have a question about quantitative risk analysis. But not everyone is as privileged to have personal access to their own risk guru—which is why this book is so welcome.

There are hundreds of books about risk management, covering almost every conceivable aspect of this fascinating topic. There are even a few that deal with quantitative risk analysis in projects. What has been lacking is a comprehensive treatment of how to actually apply the statistical techniques of Monte Carlo analysis in a way that combines intellectual rigour with practical realism. Knowing his undoubted expertise in this area, David's friends have long encouraged him to fill the gap with his own book, and at last he's acceded to the demand. This first book addresses a perennial problem faced by anyone working with projects: How can we know how long the project will take? Advances in estimating techniques have helped, as have developments on the software front. But still the question remains about how to deal with uncertainty in the project schedule, arising from ambiguity in the requirement, variability in the base estimates, and the possibility of unplanned events and circumstances (risks). This is the realm of schedule risk analysis, and the topic of this important book.

The value of quantitative risk analysis has been undermined in the past by poor practice in this area. People have performed inadequate analyses on unreliable baseline schedules using suspect input data, and then complained that the results are unrealistic. Those of us who know the power of quantitative risk analysis find this intensely frustrating! If you don't do it properly then of course the results will be of little use, and they are likely to be misleading. But the answer to misuse is not non-use, it is proper use. As befits a book intended for both the novice and the experienced practitioner, David starts at the beginning and emphasizes the importance of a good baseline project schedule. There is of course no point in trying to model the effects of ambiguity, variability and risk on a plan in which we have little confidence.

After establishing this prerequisite, David moves on to one of his favourite topics: data quality. This has always been one of David's hobby-horses, and some feel he over-emphasises this aspect. I disagree, since ensuring high-quality data is the only way to counter the first half of the infamous GIGO problem. Garbage In Gospel Out describes the all-too-common situation where people pay insufficient attention to the inputs of their risk models, and then they give too much credibility to the outputs. Instead we must use the best available data, and still be prepared to ask whether the results make sense. Recognising this problem, David provides practical guidance to the risk analyst to help him or her obtain the best possible data that properly reflect the uncertainty inherent in the schedule as well as the associated risks that might affect it. This is informed by a clear exposition of the various psychological factors that influence good estimating, including heuristics, motivational and cognitive biases.

Another unique aspect of this book is the inclusion of a novel way to take account of risks in schedule risk analysis models. David calls this the 'Risk Driver' method, and he presents it as a powerful way of mapping risks to activities. He and I have debated the extent to which this is truly new, since it has always been possible to implement this type of approach in risk models. But there is no doubt that the 'Risk Driver' method will certainly be new to most readers, and it offers a powerful way to ensure that risk is properly reflected in the risk model, as well as overcoming some potential limitations in the traditional approach to quantitative schedule risk analysis.

Other parts of this book cover parts of the analytical challenge which are essential components of any good risk model, including dealing with merge bias, and the use of stochastic branches and correlation. Most risk specialists agree that risk models cannot be realistic without incorporating such features, but they are rarely used in practice. As in the rest of the book, David's guidance in these areas is clear and helpful, explaining both why they are important and how to use them.

Like many others, I've been waiting for this book for years. I know readers will benefit from David's technical insights and practical wisdom in applying quantitative risk analysis to project schedules, and we all look forward to reading more from David in future.

Dr David Hillson, The Risk Doctor
Petersfield, Hampshire, UK

CHAPTER 1

Why Schedule Risk Analysis? Looking Beyond the Critical Path Method

Project Scheduling is Required for Good Project Management

Project scheduling is a required tool of project management. Every project must be managed to a schedule (PMI 2004). Most large projects employ project schedulers and even small projects have a schedule of some sort. The schedule predicts the completion and milestone dates of the project. It is also used to manage daily activities and resources and to record status. Scheduling has been around a long time and is one of the most widely practiced and accepted project management disciplines (Archibald and Villoria 1967).

Project scheduling is perhaps the most common discipline of project planning and control. Scheduling is practiced by thousands of individuals, with more or less training and experience, in every industry, country and organizational arena. Scheduling software abounds and the sponsoring companies are more than happy to train schedulers in their newest versions.

Projects Often Slip their Schedules

With all of this practice in project scheduling, one would think that all projects are well-scheduled and that they finish on time. However, experience tells us that projects often, even usually, overrun their schedules.[1] The degree of schedule 'slippage' is often dramatic, sometimes adding more time than has elapsed since the last status review. For instance, a real project recently added two months between monthly schedule reviews. At other times, or on other projects, schedule slippage occurs in small increments but accumulates over time resulting in substantial time delays.

Schedule delays cause problems for the project owners and contractors. Delay claims for equitable adjustments can amount to millions of dollars. Commercial windows of opportunity are missed, or made only with products that are below-standard and with incomplete functionality. Delays in completing continuous process industry projects, such as oil well or platform projects or integrated circuit fabrication lines, can cause millions of dollars of foregone income per day. Satellite lift-offs are often planned well in advance and

---

1   It is possible that since the author is involved with analyzing project schedule risk he is exposed to a biased sample of projects, mostly those that are in some kind of trouble meeting their time commitments.

if the payload is not ready it may miss the launch or launch as an 'experiment' instead of the fully functioning item. Almost all experienced project personnel have slipped a schedule at some time in their careers and some have had most of their projects slip. This is common knowledge.

In addition to slippage of the overall schedule or major milestones, the critical path often changes during project execution. In the baseline schedule a critical path is identified using typical Critical Path Method (CPM)[2] scheduling. At the end of the project, however, the path that ultimately delays the project may not be the critical path identified in the baseline, but another one. The critical path changes are often the result of some risks that have occurred to cause a previously slack path (path with positive total float) to become critical. The common experience of project personnel is that CPM scheduling does not always reliably identify the path that ultimately determines the project's completion date.

Project schedules slip and critical paths change. These are two reasons project managers establish the common practice of statusing the schedule—inputting changes that have occurred since the last time—on a frequent basis, often weekly, throughout the project.

Slippage and changes in the critical path also lead to the practice of project risk analysis. Risk analysis can address these two weaknesses in CPM scheduling.

What Causes Weaknesses of CPM Project Scheduling?

Schedules slip because of at least four problems that are commonly associated with project scheduling (Cashman 1995):

- Project scheduling is difficult. It is clear to anyone who has dealt with projects that scheduling is a demanding discipline that is not mastered by all those who are assigned scheduling responsibility.
- The rules of project scheduling—usually referring to the use of logic, constraints, resources, calendars and activity durations—are not always clear to the scheduler, nor are they followed in practice. Poor scheduling practice can lead to imperfect and sometimes dangerous project schedules. Most project schedules, even those produced by experienced schedulers, need to be debugged before they can be considered professionally competent.
- Management or the project's owner will often insist on unrealistic deadlines. The scheduler often is not permitted to produce the schedule that can be accomplished with the resources at hand. Professionalism in project scheduling would argue against many of the schedules imposed on projects today.
- Project schedules are built using single-point (deterministic) estimates of activity durations. When the uncertainty of future durations is taken into account, the duration of a schedule path is likely to differ from that computed by CPM scheduling and by CPM scheduling software.

---

2   Some information about CPM scheduling techniques that are crucial to the success of project scheduling and the conduct of schedule risk analysis is presented in Chapter 4.

Project Scheduler's Job is Difficult

Project scheduling is a demanding and exacting process. The scheduler has to keep many activities, resources, calendars, logical relationships and constraints correctly, completely and currently stated in the schedule. A scheduler is typically assigned to gather from the project teams all of the team schedules and combine them with the correct logic into one overall integrated project schedule and computer file. The scheduler must interact with both the team members and management to make sure that the schedule correctly represents the current plan and any changes therein. This means in particular that activity durations, logical relationships and the resulting critical path and total float values are examined for their realism.

The scheduler then sets a baseline schedule and begins to update (status) the current schedule on a regular basis. The scheduler may also be asked to help plan the project with 'what if' scenarios, examining alternative approaches to the project to see which will best serve the project's objectives. All of these are, among other duties, scheduler tasks.

Schedulers must have an encyclopaedic knowledge of the details of the project. They need to know more about the project than the project manager does, who often turns to the scheduler for those details at meetings with the project staff or the customer. The scheduler also must be able to communicate with the teams and their team leaders on the one hand and with management on the other. It can be argued that the technical side of project scheduling is the easiest part of the job. The majority of skills a successful scheduler exhibits involve expertly and effectively dealing with people who do not always agree that the schedule has importance and who sometimes have quite different agendas.

Often insufficient numbers of experienced schedulers are assigned. One scheduler cannot handle the entire schedule of a large project.[3] There are rules-of-thumb about the number of activities, or of open activities, that a single scheduler can handle. Anyone with scheduling or project experience has observed scheduling functions that were under-staffed or staffed with inexpert schedulers. It is no wonder that the quality of the schedule that emerges from such a situation is not the highest.

The Rules of Scheduling are Not Always Clear to the Scheduler[4]

Many schedulers are self-taught or have been initiated into project scheduling by other schedulers who have learned in a hit-or-miss fashion on the job. Many have opened the package of a popular scheduling software product and set out to produce a picture of the project timeline. Without understanding the requirements of logic that underlies a good schedule, the picture may be good on the surface but require manual adjustment each time a fact about the project changes.[5]

---

3   One scheduler actually got a deep vein thrombosis (DVT or blood clot) in his knee from sitting at his computer performing scheduling duties day and night. DVT is usually associated with lengthy airplane flights where the legs are kept in the same bent position for hours on end—this is the scheduler's plight as well.

4   See Chapter 4 for a more detailed discussion of project scheduling.

5   A number of schedule pictures have been produced in graphics packages such as Visio®, Corel Draw® and even PowerPoint®. These are not schedules but pictures to hang on the wall to impress visitors about the success of the project.

There are many excellent scheduling standards (PMI 2007), manuals and project scheduling classes available, but schedulers may not have time to attend while their project is underway. Even if they attend such classes, the messages imparted from a scheduling class may not carry over to practice on the real project schedule. Often class attendees will listen and nod their heads during the class, but then in the heat of scheduling they will return to their self-taught ways or ways they have been taught by people who are not scheduling experts. Classroom training does not impart understanding as well as mentoring does on a real project schedule by a scheduling expert.

Management May Require Unrealistic Project Completion Dates

Project schedulers should view themselves as professionals with the responsibility to produce a realistic schedule for the project. Given the responsibility and authority, most project schedulers will produce realistic, competent schedules. A dedication to applying correct best practice project scheduling discipline, when joined with good interpersonal skills, usually leads to a realistic project schedule.

Unfortunately, the environment within which the project is conducted often does not permit proper scheduling. The project sponsor or owner usually wants the project to finish sooner than realistically possible and rejects the realistic date of the project team as being unacceptable. Contractor management may also require a shortened schedule in order to please the customer or to get the bid. Many contractors will propose schedules that are consistent with owner requirements without much regard to whether it is feasible—they count on changes and other events to bail them out of a tight, infeasible schedule. Hence, because schedulers are often required by the customer, management or other important stakeholders to produce schedules that are not feasible or realistic, they will compress the schedule by overlapping parallel activities that should be done in series and shrinking durations below values that are prudent, realistic or even feasible. Faced with such pressure, schedulers will often put together a 'magic schedule' that has little possibility of success and little credibility among the project teams. Schedules such as these are very risky and quite likely to overrun.

Even in an Uncertain World We Need to Estimate Activity Duration

CPM scheduling tools, including both manual and software-based systems, are generally unable to handle the uncertainty that exists in the real world concerning duration of the project activities. This is because these tools assume that the activity durations are deterministic, that is they are known with certainty as single-point numbers. CPM scheduling methods employ simple arithmetic to add activity durations according to the project logic and derive schedule path durations. These packages also compare the exact dates calculated for parallel paths to determine the date of a merge point of those paths.

None of these arithmetic calculations using numbers assumed to be correct admits vagueness or uncertainty in the durations. Durations in the schedule must be precise, even in an imprecise world. Scheduling software packages, and manual methods, usually require the scheduler to supply activity durations, that is, the number of days the activity *will* take to complete. These duration estimates (for example, 'Activity ENG0230 *will take 17 days'*) are used by scheduling software as if they are known with certainty. In this way, as an example, durations are added when activities are in a path of activities linked with finish-to-start logic to calculate the duration of a schedule path.

The estimates of activity duration make a curious pilgrimage from uncertain estimates to rock-solid commitments. They are at first the product of duration estimating methods used by the organization. In this estimating process certain assumptions are made about resources, productivity, external events, management and the work to be done. These assumptions may be well- or ill-founded, but they soon become set and are used to calculate durations. Often the assumptions are then forgotten. The schedule dates are presumed to be exactly correct, rather than conditional statements based on the correctness of the assumptions. The durations, or at least their implied results for milestone and project completion, become targets and commitments of the performers, project teams and even team members.

Targets and commitments for completing an activity within a specific number of days lead to serious efforts to be successful and punishment for overrunning the schedule. Durations, that at the beginning of the planning process are understood as estimates, evolve into something more definite: commitments and how long the activities will really take. People who would agree at the outset that these estimates cannot be perfect often end up claiming that the estimates are engraved in stone, that 'we have no choice'.

Project Schedule Risk Analysis: Another way to Determine when Projects Finish

With all of the experience overrunning CPM schedules, we have to ask some questions:

- Can these overruns be predicted for specific projects with some degree of accuracy?
- Can the causes of project risk be identified before they become problems, possibly enhancing the project management's ability to forestall their occurrence by effective risk mitigation?
- Can we determine the causes for projects' often overrunning their initial schedules?

The answer to these questions is: 'Yes. The method is practical schedule risk analysis.' Subsequent chapters develop this message, that schedule risk analysis assists us in answering questions that CPM scheduling cannot answer (Hulett 1995, 1996; Hulett and Whitehead 2007).

Project risk analysis addresses head-on the fact that we do not know how long the activities will take. Even if an activity has been done before in a prior project, there is no assurance that it will take the same amount of time in a new project. Nor is there any assurance that adjusting it for the circumstances (for example, size, complexity) of the new project will be accurate. No two activities are exactly the same between projects. The situation will be different in the new project, with different resources and/or different productivity to be expected. Management may be different. The work may really differ from the earlier example. And, importantly, risks in the work that occurred in the prior project may not be exactly duplicated in the new project, or risks that did not occur in that prior project may occur in the new one.

In fact, it is generally impossible to know with certainty how long each activity will take or which risks will occur. Duration estimates are just that, estimates. The activities will occur in the future and 'there are no facts about the future' (USDOE 1977).[6] Even with the complete absence of management or customer interference and the use of the best estimating techniques, the actual activity durations will differ, sometimes dramatically, from those planned and included in the schedule.

Chapter Summary

Success in managing a project requires a complete and realistic project schedule that represents the project plan. Project scheduling is one of the most important skills one needs on the project team. Yet, projects often overrun their scheduled completion date. Why is that?

Scheduling is a difficult discipline and individuals thrust into scheduling are not necessarily suited to its demands nor are they always effectively trained and supported. Also, in many instances, project schedulers are not permitted to develop realistic schedules since their management and the competitive customer-contractor environment lead to optimistic—sometimes magical—schedules. These schedules sometimes lead to late delivery.

One of the most important issues facing project scheduling is the inability to incorporate uncertainty of activity durations into the typical CPM schedule. Activity duration estimates are of necessity based on assumptions that may prove untrue in fact, and the durations will differ from those estimated in many cases.

Schedule risk analysis at its most fundamental will allow us to investigate the uncertainty in activity durations and to derive their implications for the project schedule. We will be able to answer questions not possible in traditional CPM scheduling. These include:

- How likely are we to make our schedule dates?
- How much contingency time do we need to provide the degree of certainty acceptable to our organization?
- Where is the greatest risk in the schedule?

In subsequent chapters the practice of schedule risk analysis is developed from the beginning concepts and building blocks to the most sophisticated aspects of the commercial risk analysis tools now at our disposal.

---

6   Quotation is from the report's foreword written by Dr. Lincoln Moses, Administrator of the USDOE Energy Information Administration and a Professor of Statistics at Stanford University.

References

Archibald, R. D. and Villoria, R. L. (1967). *Network-Based Management Systems*. New York, John Wiley & Sons, Inc.

Cashman, W. (1995). Why Schedules Slip: Actual Reasons for Schedule Problems Across Large Air Force System Development Efforts. *AFIT Masters Thesis*. Wright-Patterson AFB, Ohio, Air Force Institute of Technology.

Hulett, D. (1995). 'Project Schedule Risk Assessment.' *Project Management Journal* XXVI(1): 21–31.

Hulett, D. (1996). 'Schedule Risk Analysis Simplified.' *PM Network*: 23–30.

Hulett, D. and Whitehead, W. (2007). Using the Risk Register in Schedule Risk Analysis with Monte Carlo Simulation. *2nd Annual Oil & Gas Project Risk Management Conference*. Kuala Lumpur, Malaysia, October 29, 2007.

PMI (2004). *A Guide to the Project Management Body of Knowledge*. Newtown Square, PA, Project Management Institute.

PMI (2007). *Practice Standard for Project Scheduling*. Newtown Square, PA, Project Management Institute.

USDOE (1977). Annual Report to Congress 1977. Washington, DC, United States Department of Energy, Energy Information Administration. 3.

CHAPTER 2 *Uncertainty in Activity Durations: Using Probability Distributions*

The Problem with Treating Activity Durations as Certain

Project milestone and completion dates, as well as critical paths, are 'determined' by looking at the results of traditional critical path method (CPM) schedules. The scheduler will develop or update (status) the schedule, calculate the dates on which the milestones will occur and identify the longest path through the project. These dates are specific and definitive. They can be stipulated in reports to management and customers. They can be laser printed.

Commitment to a schedule is one thing that all project managers and customers value highly. Performing the activities according to the estimated durations is the goal. Project teams are justifiably praised for sticking to the schedule and criticized, even punished, for overrunning the schedule. For these commitments, performances, rewards and punishments to be valid and warranted the schedule needs to be realistic.

The problem is that none of these schedule dates may be realistic or even achievable. Worse, the critical path determined at the beginning by CPM scheduling may not be the project's critical path when risk is taken into account. The definiteness of the completion dates is alluring but, in many cases, deceiving.

People take the CPM scheduling results as fact rather than projections or estimates of future events. This interpretation of the outputs of CPM scheduling is based on a belief that the logic and durations in the project schedule are known with certainty, when in fact they are more realistically thought of as estimates of, or probabilistic statements about, future events.

Reality intrudes on this happy but largely hope-driven story. Most experienced project managers and schedulers have experienced changes in the scheduled dates, usually resulting in later dates compared to those in the initial baseline schedule. Project managers recognize reality in the project execution by periodically statusing the current schedule, often on a weekly basis, because 'things change.' Often in these meetings 'slips' in the schedule are reported and there is nothing that can be done: there is no way to recover from the delays and they are just recognized and incorporated into the schedule.

Yet, in the face of these common findings, each baseline plan, and every plan that succeeds that as the project proceeds, is relied on as fact. [1]

---

1   One would think that schedulers would add contingency days to their schedule as a provision for their personal history of overruns. It is puzzling that contingency reserves are almost required in cost estimates but frowned on in

This chapter presents a better way to look at project activity durations—as probability distributions of possible duration or estimates of days the work may take for each uncertain activity. We need to be explicit about the degree of uncertainty in those estimates. This approach is a more realistic reflection of the facts on the ground in project management.

Schedules are Usually Reported as Facts about the Future

Project schedules are called 'deterministic' which is defined as: 'Describes a system whose time evolution can be predicted exactly' (Howe 2008). Determinism in scheduling means simply that:

- the schedule is assumed to represent the plan for executing the project—that is, the logic correctly reflects the approach currently envisaged;
- the durations that are assigned to individual activities are known with certainty;
- the implication of the foregoing is that the dates computed for completion of activities, major milestones and project completion as well as the identity of the critical path are all accurate.

If the schedule is the first one created during the conceptual stage of the project it may be very inaccurate. An early schedule is often too short, indicating early completion dates that the more detailed later schedules do not support. We often abandon early schedules because they cannot take account of the complexities of the project that only become known later as planning proceeds. These early schedules are often just milestones or very high-level representations of project phases (for example, design, construction, commissioning and delivery).

Early schedules may represent the owners' target dates or hopes for the project. That they are presented in scheduling software gives them an aura of content that they do not deserve. Frequently, the customer specifies the dates and the scheduler dutifully creates a schedule that, after a lot of manipulation of logic and activity durations, agrees with those dates. Mandated schedules are the most likely to be wrong and are usually short— very seldom does management want to see a long time elapse before getting the product from the performers.

The schedule dates that are derived from the schedule and its calculation of the critical path gain importance as they are reported up the chain of command, through the performing organization to the customer. Reports are made based on specific dates. Documents are (laser) printed, making those dates seem ever more accurate. They are often 'engraved in stone,' a practice that comes back to haunt the project manger and scheduler.

As these early dates become fixed in the minds of management and customers, they should be communicated with only the amount of certainty they deserve. We would like to see project managers report a certain amount of humility about their early schedule's accuracy. Dates might be reported as a range that is wide enough to represent

---

project scheduling. We find that schedule risk is one of the factors leading to cost risk and that fundamental risks in the project can affect either time, cost or both.

the uncertainty in the information upon which the schedule is based. It would be good to see schedules at early stages represent the full range of uncertainty that exists. This range would usually indicate a greater chance for overrunning than for under-running completion dates.

Reality is often otherwise, however, and even dates arrived at by schedules early in the planning process are reported with precision beyond the project team's capability to perform.

In fact, changes to the completion dates that occur as the project proceeds are often met with alarm and recriminations by customers and other stakeholders as if the original dates had been violated on purpose. They question the professionalism and effort of the project manager and the schedulers. Many believe that such violation of schedule dates is a signal of ineptness in management or performance, in some sense verging on moral weakness. A report that shows schedules slipping is cause for great consternation and accusations between contracting parties. One customer viewed the slippage with alarm and shouted: 'How are you going to bring it back to the contract date?' (In that particular case the contract date was impossible; it had been vastly unrealistic. The contractor had 'bought in' with a low time and cost bid and was paying the price for slippages at every monthly meeting with the customer.) Certainly schedule slippage has implications for contracts between contractor and customer and it may jeopardize entry into a market for a project conducted by the ultimate user.

Activity Durations are Uncertain and Best Represented as Probability Distributions

We should remember that, as Professor Lincoln Moses, then the Administrator of the Energy Information Administration, has said: 'There are no facts about the future.' (USDOE 1977) What this means is that the activity durations included in the schedule are basically estimates and that reliance on them as 'facts' works only in the remotely possible case where everything goes according to plan. There is a better way to look at activity durations in project schedules, as estimates. Estimates about the future are best represented by probability statements. Usually probability distributions are used in such cases. This is the introduction to the first and most fundamental reality addressed by schedule risk analysis that will, nonetheless, bother traditional managers and schedulers: *activity durations we put into our schedules are uncertain.*

The best way to understand the activity durations that are included in the schedule is as *probabilistic statements of possible durations* rather than a deterministic statement about how long the future activity will take. The most accurate interpretation is that the number of days actually put into the CPM schedule for any activity may just have more likelihood of occurring than the many other possible durations.[2] This 'most likely' interpretation of durations recognizes the uncertainty in estimation. It also implies the existence of specific risks which may affect activities' actual duration. Let us look at both

---

2   In fact, for many real projects we may be lucky if the durations are, objectively at the time, most likely values. Often the schedule has optimistic durations that will be hard to meet. These may be used to 'incentivize' the project teams to work harder and faster and to punish them when they do not do so. Such durations may be called 'stretch goals.'

estimation accuracy and the impact of risk on the activity duration; actual versus planned in the schedule.

Uncertainty Because of Estimation Error

Uncertainty in duration estimation is usually greater early in the project when fewer facts are known, fewer decisions are made and less information has been finalized than will be the case later as the project takes more form and shape. Even as the project approaches the execution phase, data about activity durations will not be perfect.

Activity durations are traditionally estimated with knowledge of the work to be done, the resources likely to be made available for the task, the productivity of those resources and the degree of reliance on others in completing the task among other factors. Each of these factors is uncertain and the assumptions made are subject to errors.

Often the estimating error is stated in percentages. Although cost estimates are more clearly researched and understood than are estimates of durations, there are parallels between uncertainty in line-item costs and schedule activity durations. Early in the project the cost estimate may be viewed as a '-20 percent, +40 percent' whereas later estimates made just before execution may be viewed as more accurate (for example, -5 percent, +10 percent). These percentages indicate that the estimator believes the final value will be within the range of the number calculated as of that estimate. Usually the estimating organization or person means to imply that an actual value within the stated range is within tolerance and expectations. In some organizations the project manager has authority to approve changes that are within the bounds of the estimate's presumed error.

The narrowing of the estimate bounds, implying an increase in the estimate's accuracy, is usually attributed to gaining of more knowledge as planning and engineering proceed. This increase in relevant information is usually based on increasing detail in design and securing estimates or even bids from suppliers and subcontractors. This information can often lead to sobering up from the exuberance of the early days, and can be a dash of realistic but cold water to the estimator.

Notice that estimating error, which is expressed as percentages plus and minus from the estimate in the database at the time the estimate is made, is often asymmetrical with more possibility of increases than of decreases (for example, -20 percent, +40 percent).[3] That reflects people's experience that actual cost and duration tends to get worse (higher costs and longer durations) than initial estimations. For instance, 'cost growth' is a well-researched, calibrated phenomenon whereas 'cost reduction' is often viewed as the happy result of hard work and good luck and not to be relied on. There is less information on schedule growth over time, but individual experience indicates that it would be a good thing to suspect a greater likelihood for schedules to lengthen than to shorten as the project progresses from planning to execution and finally to completion.

---

3   It is possible that estimating errors are viewed as being symmetrical—we often hear that the estimate is 'plus or minus 20 percent.'

Duration Uncertainty Because of Risk

A project risk is defined as 'an uncertain event or condition that, if it occurs, has a positive or negative effect on at least one project objective.' (PMI 2004). In this case we are thinking of the time objective and hence activity durations in project schedules. One way to examine the risks is to refer to the Risk Breakdown Structure that represents a broad sampling of areas where risks may arise to affect the project. (Hall et al. 2002; Hillson 2002, 2003; PMI 2004) A generalized risk breakdown structure looks like that shown in Figure 2.1.

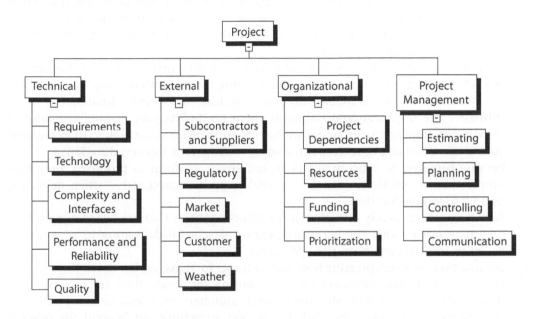

Figure 2.1   Typical risk breakdown structure

Project risks are described using a specific structured Risk Statement (Hillson 2000, 2004; Hillson and Simon 2007) with the following form:

Because of a (CAUSE), the (RISK) may occur and if it does some (IMPACT) will result. Notice that:

- The RISK is the uncertain component of this definition: it includes a word such as 'may' or 'might' and it is in the future.
- The CAUSE (for example, 'We have to develop this iron ore mine in the jungle.'), while often confused with the risk, is a fact and without it we might not have the risk. Confusion of the cause with the risk could lead to inaccurate calibration of the risk's probability and impact and to inappropriate risk responses (for example, 'Let's move the iron ore to a different site, then mine it.').
- The IMPACT is on a project objective such as cost, time, scope or quality. The statement implies that the impact is a single value, but in more recent work the risk can have a range of impacts (see Chapter 8).

Because we are looking at duration risk we need to examine risks that will have an impact on activity durations if they occur. In fact there are many risks that seem to have more impact on project durations than on project cost, scope or quality. It is easy to slip a schedule, as we find in many cases. Often we do not know why it happened, but by looking at and analyzing the risks we can do a better job of understanding, sometimes in advance, whether and by how much the schedule may slip.

Examples of risks that may lead to uncertainty in activity duration over and above the implication of estimating error (see above) can be such as:

- We are planning to use a new technology. Because we do not know the full extent of the planned technology, implementation may be more difficult (or easy) than assumed during estimation, leading to actual durations that are longer (or shorter) than those included in the schedule.
- Because of a matrix management of resources in an organization that has several projects going at once, resources assumed during the project's development may not be available or may be available only part time, leading to longer durations.
- While we use organizational standards about productivity in making estimates of activity duration, resources' actual productivity may be lower (or higher) than assumed during the schedule's development, leading to longer (or shorter) durations.[4]
- Because suppliers or subcontractors do not always perform as promised, they may supply their materials late or the materials may be unacceptable when delivered, leading to longer durations.
- Because we need to rely on external regulatory agencies that are out of our control, they may not provide the necessary permits as scheduled, leading to advances or delays in starting certain on-site operations.
- Because there is severe pressure from the customer to 'make' earlier dates, the schedule has been shortened unreasonably by adopting durations that are unreasonably short. Executing a project with these 'magic numbers' may lead to many activities overrunning their durations, with the project stretching out beyond the original schedule.

It is important to notice that many of the risks listed here, and indeed many other risks, lead from uncertainty about an assumption made during the generation of the schedule. Project schedulers and team leaders who make the estimates of duration will make assumptions whether implicitly or explicitly. There is often a good basis for most of the assumptions, although some may be imposed on the estimator by the pressure of the customer, management or the competitive situation. Whatever the source, these assumptions may or may not be listed. Ignoring the uncertainty in the assumption once a single-point duration estimate is produced (for example, design of this circuit will take 60 working days), or forgetting the various assumptions that were made will lead to schedule problems as events unfold.

Notice also that many of the risks listed have both opportunity and threat characteristics. That is, the uncertain event may, if it occurs, improve (opportunity) or harm (threat) the

---

4   Because labor productivity is always present we are thinking of a 100 percent likely risk with only the level of that productivity uncertain. This type of risk is often called 'ambiguity.'

project. This aspect agrees with modern definitions and thinking of project risks. (Hillson 2004; PMI 2004) Not all organizations would recognize an opportunity as a risk.[5]

Activity Duration is Best Represented by a Probability Distribution

Completion of an activity in a project schedule may take more or less time than is stated in the project schedule because of both estimating inaccuracy and project risk. Thus, it is useful to represent the duration, realistically, with a probability distribution. This distribution represents the various durations that may be necessary to complete the work of the activity under different sets of assumptions. Hence there is at least a shortest time the activity can take and still be completed so its successor can start (if there is a finish-to-start logical relationship) or finish (if the logic is finish-to-finish). In the limit the activity cannot take negative time, although a longer duration can probably be identified that is the practical shortest time for the work.

There is also a longest duration that this activity can take, although that value is more problematic since conceptually an activity might never be completed, perhaps representing the 'Black Swan' concept. (Taleb 2007) Putting aside that extreme possibility for the moment, there is a longest duration that may be produced by dire circumstances for the activity. The probability distribution of durations includes all of the possible durations which are no fewer than the shortest duration possible and no greater than the longest duration possible. In concept there is an infinite number of possible durations, even between an optimistic duration of 25 days and a pessimistic duration of 40 days because duration could be measured in minutes, seconds or even finer distinctions. In practice, we usually measure durations in working days.

Some of the durations between 25 and 40 days are more likely than others. A probability distribution represents the relative likelihood (or probability, depending on your preferred terminology) of each of the possible durations included. Early in the project the team may not have a fixed view that any particular duration is more likely than any other within the optimistic-to-pessimistic range. However, most probability distributions of possible durations have a single duration that is more likely than any other, if only by a little bit and if only held with very little confidence. That duration usually does not occur at the extreme optimistic or pessimistic possible value, although it could do so.

The typical probability distribution representing possible activity duration is continuous. Continuous means that the activity can take any duration within the range and that there are no discrete jumps between adjacent possible durations. Hence, very few distributions would be of the variety that an activity could take 1 week, or 2 weeks but not 6, 7, 8 or 9 working days. We will generally assume also that most duration distributions have a single most likely value and diminishing probability toward the optimistic or pessimistic duration limits.

---

5   Some organizations would classify these as uncertainties. They would classify what we are calling threats as risks and include opportunities as another type of uncertainty. This issue is not that important, although it has led to spirited debates. Whatever we call it, good or bad things can result from the uncertain event.

Finally, we will not assume that the distributions need to be symmetrical around a mid-point. A symmetrical distribution would have the highest single value in the middle of the range, equidistant from the optimistic and pessimistic extremes. Because project risks tend to be asymmetrical between optimistic and pessimistic extremes, the favorite probability distributions that seem to fit most project management schedule (and cost) risks are the Triangular and the Beta as shown in Figure 2.2 and Figure 2.3 below. These two distributions have distinct peaks at the most likely value and their probabilities

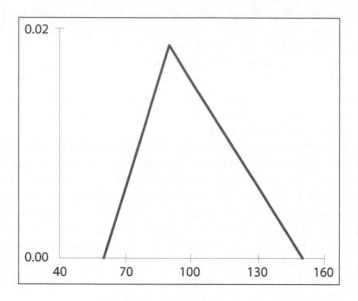

Figure 2.2   Triangular distribution I

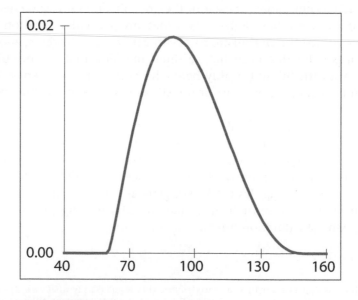

Figure 2.3   Beta distribution I

become smaller until they vanish at the extreme optimistic and pessimistic values.[6] Of course triangular and beta distribution may be symmetrical, it is just that they are not necessarily so.

The triangular distribution is uniquely and completely specified by the three-point estimate that is common among estimators, the optimistic-most likely-pessimistic triad. The sides being straight, all software will produce the same triangle if given the same three points.

The beta distribution is a little trickier than the triangular. The traditional beta is specified by a mean and standard deviation, data that usually cannot be collected from project or subject matter experts. In particular, the concept of the standard deviation is murky, even to people involved in the risk analysis profession. By good luck the software developers have developed a distribution usually called the 'BetaPERT' which adopts a beta shape but is specified by the three-point estimate. The problem with 'the' BetaPERT is that it is not unique. There is no 'the BetaPERT' and each piece of software has its own version. Usually this distribution is significantly taller and thinner than the triangular distribution for the same three points, providing the user with a distinctly different alternative choice. Of course, as we see below, these distributions should be used with an appreciation of the risks they represent.

Two other distributions are commonly available for project schedule (and cost) risk analysis. They are the Normal and the Uniform, shown below in Figures 2.4 and 2.5. The normal (Gaussian) distribution has the benefit that it is commonly found in nature. It also has no limit to its low and high range values, although practically there is insignificant probability outside of the mean plus or minus two or three standard deviations. The normal distribution has three drawbacks when considered for practical use in project risk management:

- it must be symmetrical, that is the peak must be equidistant from the optimistic and pessimistic values. For this reason, we have little use for the normal distribution in project risk analysis;
- it is specified by two parameters, the mean and the standard deviation. The second of these two, the standard deviation or 'sigma' is generally not understood by interviewees; and
- activities have an irreducible minimum duration, often known to the participants, and certainly activities cannot take negative time, whereas the normal distribution has no lower bound. All in all, one could argue that there is very little use for the normal distribution.

The uniform distribution represents the possibility that there is no duration between the optimistic and pessimistic that is more likely than any other. Circumstances, such as a lack of data in the conceptual stages of the project, can lead to this conclusion.

Of course there are many durations other than the four discussed here. Some software packages offer Lognormal, Gamma, Rayleigh, Weibull and others including non-continuous distributions such as the Discrete. Some people would like to construct their own distributions, perhaps with a bi-modal shape.

---

6   These and the next two charts were developed in @RISK®, a simulation program for Microsoft Excel®. @RISK® is a product of Palisade Corporation.

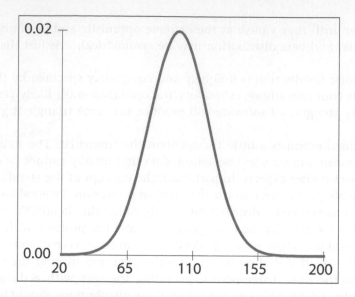

Figure 2.4   Normal distribution I

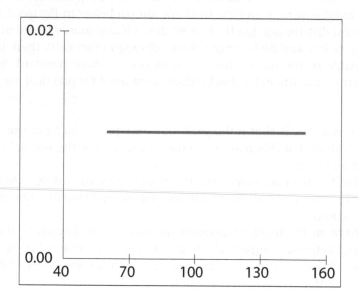

Figure 2.5   Uniform distribution I

Usually the effort expended in selecting just the right distribution is greater than the payback. It is usually the case that the effort worrying about the precise distribution shape, rather than using these common distributions, would be better spent specifying the three-point estimates. The analyst should spend time interviewing more individuals on the same activities or spending more time with the primary interviewees getting their rationale for the three points and tuning their values than worrying about the precise distribution shape.

This discussion is not to say that one shape, say the triangular, is to be used at all times. Rather, the four most common shapes seem to be sufficiently diverse to represent the variety of risks in most cases. Still, choosing between them is the analyst's responsibility.

The Three-Point Estimate

It is realistic to represent the duration of typical work-type activities with probability distributions, so we need to collect data that will represent the values (days in most cases) and their relative probability of occurring on this project for each activity with an uncertain duration. Collecting these data will allow us to input to the risk analysis program the specific distribution (type of distribution and parameters) that is appropriate for each risky activity. The data to be collected helps put boundaries and shape to the probability distribution of possible durations for each risky activity. The typical three-point estimate is made up of the following three data items.

THE LONGEST OR PESSIMISTIC DURATION

The longest duration is the result of several factors that are important in affecting its duration going 'wrong' simultaneously, if that is even remotely possible. When the duration is estimated the person conducting the risk analysis probably made some explicit or implicit assumptions. Discussing the longest duration, which is called 'pessimistic duration,' those assumptions are re-examined and other risk events are potentially included. The pessimistic duration is the result of pushing these risk assumptions to their pessimistic extremes together, if that is possible on this activity. The duration could be long if this new, pessimistic, scenario were to occur (a conditional statement). Generally, a good definition is that it is only 1 percent likely to be longer than this duration.[7]

A good question is: 'What should we do about the possibility of the Black Swan?' (Taleb 2007). The Black Swan is the highly improbable event that is, if it occurs, highly impactful. A characteristic of the Black Swan is that it is hard to predict, although after the fact there may be multiple theories about why it occurred. Many interviewees are accustomed to thinking in tunnels or stove pipes, 'inside the box,' when the Black Swan event is clearly an outlier or 'outside the box.' In general, while we try to explore this phenomenon we are not usually successful. The analyst must be alert to these possibilities and their potential impact. But, if found, the Black Swans may be posited as a scenario outside of the Monte Carlo analysis we describe in this book.

THE SHORTEST OR OPTIMISTIC DURATION

This is the result of several factors going 'right' together on the activity. It is called the 'optimistic duration' because finishing sooner is a good thing for most projects. Generally, a good definition is that it is only 1 percent likely to be shorter than this duration. This duration may represent opportunities to improve on the project schedule.

---

7   It is difficult to get someone to promise that the activity would never, ever, exceed any number of days. Since we 'never say "never"', let us give the interviewee a 1 percent chance of being wrong. Of course this value is used as the 100 percent case, not the 99 percent case, in the analysis, but this difference has little meaning when we use Monte Carlo simulation, because we seldom provide for the very extreme outliers in making promises or provisions.

THE MOST LIKELY DURATION

This is the result of the most likely scenario, the realistic scenario for that activity on that project being conducted by the organization and in the relevant period of time. This duration is viewed as more likely than any of the durations in the possible range from shortest to longest.

It is important to note that *the most likely duration need not be the duration in the schedule*. This is an important fact of schedule risk analysis. It is tempting to assume that the most likely duration has been entered into the schedule by a rational project manager and scheduler. If we ask the question of the schedulers or project teams we may be surprised to learn that the schedule duration is old, overtaken by events, superseded by more recent information, biased (usually to be shorter than it should be) by the force of management or the customer, or was just a mistake. The most likely duration in the project risk analysis is often different and may be greater than that in the schedule.

Which Distributions to Select to Represent Uncertainty in Activity Durations?

These concepts can be illustrated using a simple example of a probability distribution of durations for an activity. It is not surprising to find facts such as these during an interview:

- schedule duration: 45 days;
- optimistic duration: 35 days;
- most likely duration: 45 days;
- pessimistic duration: 70 days.

Different distributions will represent these estimates in ways that mean quite different notions about the risk in the activity's duration.

Typical distributions using different assumptions are shown below in Figures 2.6 to 2.9. The following charts represent Monte Carlo simulations of the assumption distributions, so they show how these four alternative distributions are implemented. Monte Carlo simulation is discussed in the next chapter.[8]

Comparing the distributions we can see how they represent the probability of all durations between 35 days and 70 days, that is using the same three-point estimates.[9]

First, in Figure 2.10, we can compare the triangular and the beta distributions, the two most common distributions that need not be symmetrical.[10]

Notice the differences between the triangular and beta distributions. A typical beta or BetaPERT distribution derived from the three-point estimate concentrates most of the probability around the most likely duration (45 days in this case) and implies a mean of

---

8   These charts are made with Crystal Ball®, a simulation package for Microsoft Excel. Crystal Ball is a product of Decisioneering, Inc., an Oracle company.

9   Of course the uniform distribution does not use any 'most likely' duration since each number of days is viewed as equally likely. In addition, the normal distribution typically ignores any specified most likely duration and places the peak of that distribution equidistant from the optimistic and pessimistic values.

10   The overlay charts below are outputs from Crystal Ball®, created by Decisioneering.

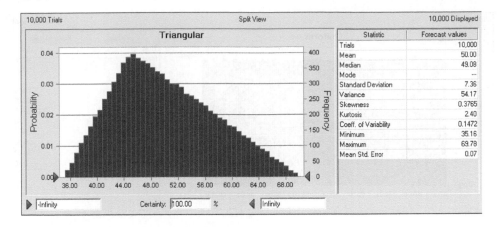

Figure 2.6   Triangular distribution II

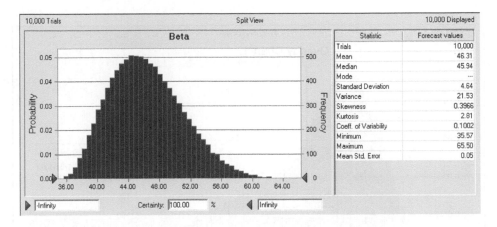

Figure 2.7   Beta distribution II

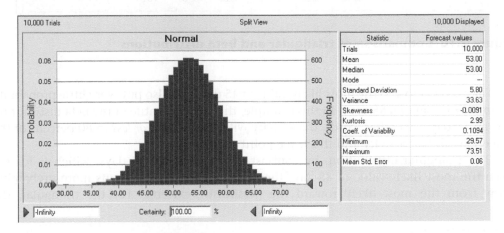

Figure 2.8   Normal distribution II

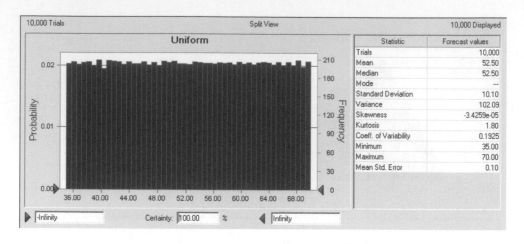

Figure 2.9   Uniform distribution II

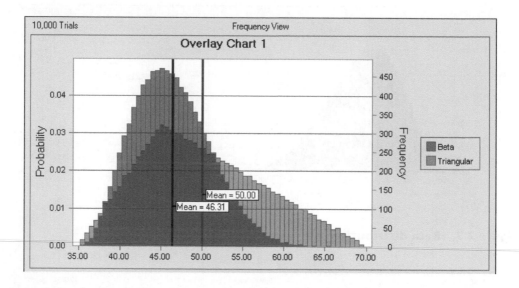

Figure 2.10 Comparison of triangular and beta distributions

46.3 days, very close to the most likely value of 45 days.[11] It also puts very little probability on durations beyond 55 days. In this example, the 80th percentile of the beta distribution is only 50 days. (In other words, the activity will take *50 days or less* in 80 percent of the projects with this activity and this three-point estimate.)

The triangular distribution, by contrast, has a mean of 50 days, which is further from the most likely value of 45 days than the beta's mean, and puts more probability away from the most likely toward the pessimistic extreme. For the triangular, the

---

11   The 'mean value' is the weighted average of all possible durations. In calculating the mean value, each possible duration is multiplied by its relative likelihood of occurring (represented by the height of the column at that day's value).

80th percentile is 57 days in this case. Comparing the 80th percentile values is common since many project managers want to provide for the bulk of the risk.

These two commonly used probability distributions are quite different from one another. The triangular distribution shows less confidence in the mean and gives more importance to the extreme values, particularly the pessimistic. The triangular distribution is used in many cases because it represents a greater amount of uncertainty in the most likely duration than the beta does. The triangular distribution is also frequently used because it is completely and accurately specified by the three-point estimate, concepts that the interviewees can grasp. One drawback with the triangular distribution is that it does not appear in nature, so strictly speaking it is at best an approximation of the population of durations that will be found.

The uniqueness and accuracy of the triangular distribution contrasts with the fuzziness of the beta distribution. There is really not one beta distribution that is consistent with the three-point estimate, even though one is shown above. The beta shown above is just one of a family of beta distributions specified by its two 'shape parameters.' Shape parameters determine the shape of the specific beta, including whether its peak is close to one extreme or the other and whether it is tall or squat. Unhappily, there is no way to interview people about their beta distribution's shape parameters. Shape parameters are less well understood than the standard deviation. In fact very few people even know about the beta distribution's shape parameters.

If the sum of the shape parameters is large the beta drawn is tall and slender and implies more confidence in the most likely duration. If the sum of the shape parameters is small, the beta drawn is short and squat, implying less confidence in the most likely duration. Two rather extreme beta distributions, consistent with the same three-point estimate but with different shape parameters, are shown in Figure 2.11.

Happily, most of the distributors of Monte Carlo simulation software have made some decisions for us, choosing a combination of the two shape parameters that forces a fairly tall and slender beta shape that is quite distinct from the triangular distribution for the

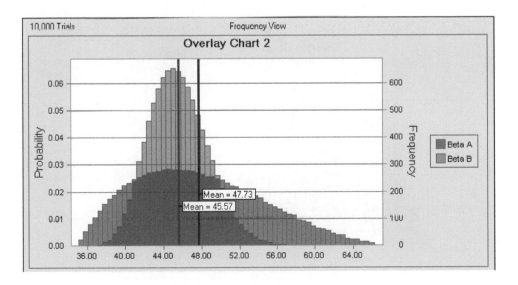

Figure 2.11 Comparison of two different beta distributions

same three-point estimate. Unhappily the user generally cannot change the software's decision to reflect more or less confidence in the most likely duration.

The normal and the uniform distributions are necessarily symmetrical. For the normal distribution, in practice that means the optimistic and pessimistic durations are the same number of days below and above the value chosen (by the software, not the analyst) to be the most likely. For the uniform distribution, all values are equally likely so there is no concept of the 'most likely.'

Because most project risk analysis distributions are asymmetrical, these distributions are of limited use in project schedule risk analysis. In the cases where a 'plus or minus the same number of days from the most likely value' is appropriate, either the beta or the triangular distributions can be symmetrical.

All four most commonly used distributions are shown in Figure 2.12. From this figure it is clear that they tell quite different stories about the uncertainty in activity durations even though they are consistent with the same three-point values. It is important that the analyst is sensitive to the story that is being told during the interviews about the risk in order to choose the most appropriate distribution. If the story is complex and a different distribution would be required, most software packages offer other distributions or ways to construct them.

One measure of the degree of certainty in the mean (not the most likely) distribution that the different distributions imply is the standard deviation, sometimes called 'sigma.' It indicates the range around the mean value within which the majority of the probability lies. A higher standard deviation implies less confidence that the activity's ultimate duration is close to the mean value of the distribution, whereas a small standard deviation implies that the ultimate duration is reliably close to the mean value. The four distributions shown in Figure 2.12 have different means and different standard deviations, as shown in Table 2.1.

It is clear that the triangular and the beta distributions are quite different, with the beta's mean closer to the mode and the beta's standard deviation only about 60 percent

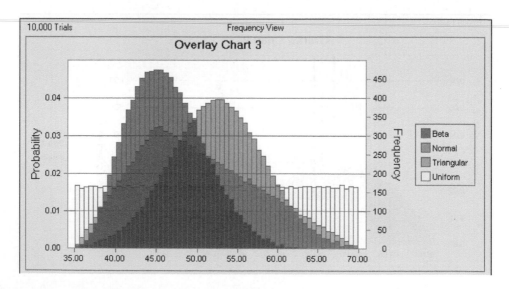

Figure 2.12 Four distributions compared (35 days, 45 days, 70 days)

Table 2.1 Comparison of four popular probability distributions

Comparison of Three Distributions			
Three-Point Estimates (35, 45, 70)			
Distribution Type	Mode	Mean	Standard Deviation
Triangular	45.0	50.0	7.4
Beta	45.0	46.3	4.6
Normal	52.5	52.5	5.8
Uniform	N/A	52.5	10.1
'Mode' is the technically correct term for the Most Likely of the text. The normal distribution is specified by the mean and the Standard Deviation. 5.8 is typical (average of low of 35 and high of 70).			

as large as that for the triangular. Many risk analysts choose the triangular by default over the beta because it represents less confidence in the mode and interviewees tend to underestimate the ranges of risk (see Chapter 5), but that may not be the case for a specific activity. For instance, the beta might be a better choice than the triangular if the interviewee specifies a very long pessimistic duration but indicates that, while possible, there is not much probability as the days approach that extreme point.

A later chapter (Chapter 5) discusses the problems encountered in collecting data mostly relying on people's judgment. In that chapter the various biases discussed must be identified and overcome. There we focus on problems identifying the three-point estimates because that is much more important than whether the probability distribution is one shape or another.

Still, there are some reasons to choose a distribution other than the four emphasized here. People may complain, for instance, that it is important to have a distribution like the Lognormal because there is, strictly speaking, no upper limit to how long an activity may take. Activities are theoretically open-ended because they may never be completed, and a lognormal has the property that it never quite hits the X-axis at its upper range. This is all well and good, but it is of very limited if any importance to the process or practical schedule (or cost) risk analysis. Extreme values with extremely low probabilities will be selected rarely in the Monte Carlo simulation, our main tool of analyzing these input data, and most people will not focus on the extreme values in the answers in any case. Hence, suppose there is an outlier that is theoretically possible, and if that outlier is chosen, say once in 1000 trials, extending the schedule by, say, a year. The probability of that extreme event is so small that the customer usually does not focus on that value when choosing the date to promise the project will be complete. Many users of risk analysis results focus on resulting dates that are between 50 percent and 80 percent likely to be adequate for the project completion. A very rare event, while possible and potentially disastrous, will not be relevant to the majority of project stakeholders.

Summary of Using Probability Distributions to Represent Duration Uncertainty

This chapter deals with the essential truth that durations of the activities in a project schedule, being in the future and estimated by humans, are not known with certainty. It presents a more realistic way to look at activity durations, as probability distributions of possible duration. Four commonly used distributions are presented and compared.

The typical and traditional CPM schedule is based on a deterministic view of the world including that the activity durations are known with certainty. If this were true, a completion date and critical path could be calculated with certainty. This happy circumstance occurs only, if at all, when things go according to plan. Since this situation never or hardly ever occurs, the dates of the CPM schedules are of limited value except as a target or as a starting point for a schedule risk analysis.

Most experienced project personnel know the problems that occur when we rely on the CPM schedule dates, and they expect the final result to be different. Even though they are committed to making the schedule, these experienced people recognize the possibility of change by scheduling periodic schedule update meetings and reissuing the schedule with new dates thereafter. However, with all of this experience, a slip in the schedule is often viewed by management and customers with alarm and recriminations as well as demands for recovery and penalties. The possibility of schedule slips is often not discussed in advance and their possibility is often officially denied. Yet most projects overrun their original or baseline schedules. Schedule risk analysis helps to gain an understanding and to calibrate the possibility of schedule results other than those baselined.

The essential issue in practical schedule risk analysis is that activity durations are not known with certainty but are estimates of what may occur in the future. They are based on assumptions that may not come true about resources, productivity, reliance on external parties and other factors. Often customers and management impose on the schedule unreasonable completion dates, and activity durations have to be shortened to meet those dates, leading to unrealistic schedules and overruns.

This chapter has introduced the concept of the three-point estimate, made up of the optimistic or shortest duration, the most likely or most commonly found duration, and the pessimistic or longest duration. These are concepts and values that can be collected from project participants, mainly in workshops or interviews. These interviews can take 1–2 hours or more depending on the number of data items to be collected, the experience of the interviewees in risk data gathering, the skill of the interviewer and the depth of the discussion concerning causes of the ranges. The interviews need to start with the risks of the project and we introduced the risk breakdown structure to illustrate the breadth of risk concerns that should be considered.

We use Monte Carlo simulation (described in the next chapter) as the primary tool of practical schedule risk analysis. Simulation requires software tools and several of these have been developed that are used with popular scheduling packages. Several different probability distributions are typically offered by manufacturers of these programs. These distributions are discussed in this chapter. The triangular and beta distributions are common choices.

Each of these two probability distributions has the appealing property that they may be asymmetrical, matching the typical risk found in project management. The triangular can be fully specified by the three-point estimate while the beta needs a further assumption, which is sometimes hidden from the user, about two shape parameters. The normal and uniform distributions can also be used, although they must be symmetrical and hence have limitations for representing typical project risks. Other distributions may be available, but it is probable that they will not be needed. Excessive attention to the precise shape of the distribution is viewed as unproductive. If time is available, a more productive use of it would be to develop more accurate and understood three-point estimates.

References

Hall, D. et al. (2002). The Universal Risk Project Final Report. INCOSE PMI.

Hillson, D. (2000). 'Project Risks – Identifying Causes, Risks and Effects.' *PM Network* 14(9).

Hillson, D. (2002). The Risk Breakdown Structure (RBS) as an Aid to Effective Risk Management. *5th European Project Management Conference*. Cannes, Project Management Institute.

Hillson, D. (2003). 'Using a Risk Breakdown Structure in Project Management.' *Journal of Facilities Management* 2(1): 85–97.

Hillson, D. (2004). *Effective Opportunity Management for Projects: Exploiting Positive Risk*. New York, Marcel Dekker, Inc.

Hillson, D. and Simon, P. (2007). *Practical Project Risk Management: The ATOM Methodology*. Vienna, VA, Management Concepts.

Howe, D. (2008). 'The Free On-line Dictionary of Computing.' http://dictionary.reference.com/browse/deterministic.

PMI (2004). *A Guide to the Project Management Body of Knowledge*. Newtown Square, PA, Project Management Institute.

Taleb, N. N. (2007). *The Black Swan: The Impact of the Highly Improbable*. New York, Random House.

USDOE (1977). Annual Report to Congress 1977. Washington, DC, United States Department of Energy, Energy Information Administration. 3.

3 Uncertainty Along a Schedule Path: Using Monte Carlo Simulation

Scheduling if We Know the Activity Durations

In traditional deterministic project scheduling using manual or automated scheduling techniques the activity durations are assumed to be known with certainty. Under that unrealistic assumption the total duration of a path can be determined by simple mathematics of adding durations, assuming finish-to-start logic and following the logical relationship of the activities. The completion date is known and reported up to management.

Scheduling in Real Life When the Durations are Just Estimates

After reporting to management 'the schedule' results, wise schedulers will cross their fingers, at least figuratively, knowing that many things can happen to their estimates[1] and also knowing how the estimates rely on assumptions that have yet to be validated.

We have seen that activity durations that appear in project schedules are basically estimates of how long the work of each activity will take in the future. These are projections, sometimes carefully constructed but other times closer to guesses. Many factors may cause the actual duration to differ from the estimate, including external factors, internal inefficiencies, poor estimates to begin with and assumptions made to support the estimates that may be proven wrong. The best way to represent the reality that actual durations may, and probably will, differ from the baseline estimate is to show the activity durations as probability distributions. Several commonly used distributions were presented in the previous chapter.

But how can we determine the duration of the path if we do not know how long the activities will really take? Using this approach, activity durations are represented by shapes or distributions and *we cannot sum shapes*. The challenge is to determine the risk of paths of activities when the activities' durations are not known. We need to calculate the path uncertainty represented by combining, not summing, all of the path's activities' probability distributions.

---

1   Changes can occur with the schedule logic as well, but we will focus on uncertainty in duration of the current plan represented by the schedule logic. Only in later chapters (see Chapters 7 and 9) will we introduce the possibility of changes to the plan.

Since the durations are represented by probability distributions, we need to find a method that will allow us to combine distributions down the path, respecting both the duration uncertainty and the logical relationships between the activities.

The accepted and most commonly available method used to combine probability distributions is called Monte Carlo simulation. The application of Monte Carlo simulation methods to analyzing project schedule risk is the subject of this chapter.

What is a Schedule Path?

A schedule is not made up only of activities, however, but of strings of activities linked together in logical sequences. The logic represents the current plan of how the project will be accomplished. Some activities have to be performed before others, while some activities can be worked in parallel. Four main logical relationships between activities are offered in scheduling software:

- Some activities (for example, constructing the framework of a house) may not start until a predecessor (for example, curing of the concrete house foundation) has finished. Sometimes several predecessors (for example, building various subsystems for a satellite payload) must have finished before the successor (for example, integration of the subsystems) can start. These activities are linked by the common (default) finish-to-start (F-S) relationship. Finish-start logic is the most common logical relationship at the detailed level of project scheduling where activities are short and limited in definition.
- Other activities that are related may be able to occur simultaneously and are planned to be executed in parallel. Sometimes one activity starts (for example, design of components) and, after a while ('a while' is represented by a lag of a few days, weeks or months) the successor activity (for example, drafting of the sheets needed for fabrication or construction in the field) can start. This is the start-to-start (S-S) relationship.
- Another logical relationship linking simultaneous activities is the finish-to-finish (F-F) relationship. For instance, the drafting team in the paragraph above cannot actually finish drafting until the final designs are produced. Also, that last drafting work may take some time, perhaps a few days. Hence, drafting cannot finish until some number of days, represented by another lag, after design finishes.
- Start-to-start and finish-to-finish relationships occur frequently in summary or early schedules when the details are not known and the work is represented by summary work or phases.
- Finally, there is the rarely used start-to-finish logic. This means that a logical successor (for example, software coding) will not finish until its logical predecessor (for example, printing of the technical manual) starts. Often inexperienced schedulers use the start-to-finish logic inappropriately instead of the standard finish-to-start. Sometimes scheduling is planned from the finish date backward to the start date and this relationship is used.

In Chapter 4 'A good project schedule is needed: Critical Path Method scheduling 101,' several issues about schedule logic are discussed. In this chapter we will examine schedules with the simplest of the logical relationships, finish-to-start, so we can concentrate on the methods and issues of determining the risk in schedule paths based on risk represented by probability distributions of activity durations.

Combining Activity Durations along a Schedule Path

A seemingly straightforward schedule with simple finish-to-start logic might look like Figure 3.1.

This project starts on June 1 of some year. Notice that the first activity, Design, which takes 30 days, finishes on June 30, so we do not have to worry about weekends. (For convenience we assume a 7-day calendar.)

Using the Critical Path Method (CPM) of scheduling we see that this is a 115-day project ending on September 23. This duration and date are determined by simply adding the durations (30d + 50d + 25d + 10d = 115d) and applying that duration to a calendar. Every scheduling software package will show the same result.

As we have shown in the previous chapter, however, we cannot predict with certainty that Design will take 30 days, Build will take 50 days, and so forth down the path. There is no doubt that the project team is committed to this schedule and that the customer is expecting to receive it on September 23. There is probably a contract between the project performing organization and the customer that includes these dates. But there is no certainty in these numbers. Can we do better at estimating the completion date?

A prudent contractor, and indeed a prudent customer, might (but often does not) ask: 'What is the possibility that this project will not deliver on September 23?' This question is important if delay in the receipt of the article precipitates a contract claim for non-performance or is important to the beginning of another activity. For instance, if a wedding dress is supposed to be altered and delivered on September 23 for a wedding on the morning of September 24, it does no good at all if it is even 1 day late.

ID	Task Name	Duration	Start	Finish	May	June	July	August	September	October
1	**Simple Project**	**115 d**	**6/1**	**9/23**						
2	Start	0 d	6/1	6/1		6/1				
3	Design	30 d	6/1	6/30						
4	Build	50 d	7/1	8/19						
5	Test	25 d	8/20	9/13						
6	Deliver	10 d	9/14	9/23						
7	Finish	0 d	9/23	9/23						9/23

Figure 3.1   Simple one-path schedule

Combining Activity Durations along a Schedule Path using Simulation

The answer to whether the project will be finished on time leads to schedule risk analysis. In order to determine the risk that a series of activities can lead to any particular date we need to accommodate the uncertainty of each of those activities' durations in the correct way. There are two key steps to this analysis:

1. Evaluate the risk of each of the activities and represent their durations with probability distributions. This process was discussed in Chapter 2.
2. Combine the probability distributions using Monte Carlo simulation to derive the probability of the final date's occurring.

By computing the probability distribution of the completion date for this simple four-activity one-path project we can answer some questions that CPM scheduling cannot answer:

*   How likely is it that this project will finish on the completion date indicated in the project schedule, in other words, how likely is it to be finished on time? For this simple project, with the risks we have on each activity, how likely is it to finish on or before September 23? Since the start date is fixed in our example, this question is equivalent to finding the probability distribution of the total duration, for example, 'Is completion likely to be in 115 days or fewer?'
*   How much time contingency (extra days added to the schedule) do we need to establish so that we can say with a desired amount of confidence that we can finish by that date? On this project, how many days do we need to add to provide a date with, say, a prudent 80 percent level of confidence that it will be finished on that date or earlier?

Collecting the Data for Risk

First we determine the risk of each individual activity. This is done by interviewing knowledgeable experts in the fields of design, build, test and delivery. Often these people are involved in the process of planning this project. Direct participants and those responsible for this particular project may be biased to show that it will be successful, so we may want to get information from other experts not involved directly in this project as well.

The main data we collect is the three-point estimate, as described in Chapter 2, 'Uncertainty in activity durations: using probability distributions', which is a standard and practical way to represent the risk. As discussed in the prior chapter, these three points are:

1. Optimistic (because we usually want earlier delivery) or the very shortest possible duration for the activity on this project.
2. Most likely duration, the duration for the activity that will occur most often in repeated conducts of the same project (without learning effects).
3. Pessimistic (because we usually want to avoid late delivery) or very longest duration for the activity on this project.

There are some biases in data collection, and these are discussed in more detail in Chapter 5, 'Collecting risk data: exploring methods and problems.' In this chapter we assume that the data collected represent the risks realistically.

A possible result of risk interviews on these activities in our simple sample project is shown in Table 3.1. It indicates that there is more risk of overrun than of under-run on Design, Build and Test activities but the Deliver activity is equally likely to overrun as under-run. It shows that the Test activity has the most risk of overrun. Also, the interviews confirmed that the durations shown in the schedule are still viewed as most likely. Finally, the risk of overrunning, while present, is not extreme, except perhaps for the Test activity.

Since the activities are logically linked finish-to-start, the simplest thing to do would be to add up the columns and examine the risk under each scenario. The temptation to add the columns, as shown in Table 3.2, to find out how long or short the project may be, *should be vigorously avoided*. The sum of the columns would represent the following three scenarios:

1. Optimistic, where all activities are as short as possible (this is wildly optimistic—it will not happen);
2. Most likely, where all activities take their most likely duration (this scenario is unlikely to happen);
3. Pessimistic, where all activities are as long as possible (this is wildly pessimistic—it will not happen).

Table 3.1   Example three-point estimates of activity duration

Example of Activity Duration Risk from Interviews		
Activity	Optimistic	Most Likely	Pessimistic
	(Working Days)		
Design	20	30	45
Build	40	50	65
Test	20	25	50
Deliver	5	10	15

Table 3.2   Calculations of all-optimistic and all-pessimistic results are misleading and must not be made or reported

Calculations that are NOT RECOMMENDED And May Damage Project Understanding		
Activity	Optimistic	Most Likely	Pessimistic
	(Working Days)		
Design	20	30	45
Build	40	50	65
Test	20	25	50
Deliver	5	10	15
TOTAL	85	115	175

People often like to see the sum of the optimistic values or the sum of the pessimistic values 'just to get an idea of the project risk.' The idea they get by doing this exercise is misleading, giving both too much hope or optimism (this project will not take only 85 days) and too much to fear from pervasive pessimism (this project will not take as many as 175 days). Yet, people persist in wanting to do these calculations. (A recommended alternative is shown in Figure 3.7.)

What is the likelihood that this project will take only 85 days? Each of the four activities has to be finished in record time (remember, there is only a 1 percent chance that the durations will be shorter than the optimistic duration, by definition) and the hand-off from one activity to the next must be flawless. Most project managers and others with experience will report that no project (one can never say 'no project' but so far none has stepped forward to make this claim) finishes with each and every activity performing to their utmost capacity and finishing as early as possible. While we hear that 'everything has to go perfectly to achieve this schedule' that is usually a code for 'do not believe this schedule—it'll never happen.'

What is the possibility that this project will take 175 days? While some skeptics may suggest that this scenario is somewhat more likely than the 85-day one, this too is very unlikely. It is only remotely possible that each activity will take the longest that it could possibly take, together on this one project, although it is not technically impossible (remember that, by definition, the actual duration is supposed to be longer than the pessimistic in only 1 percent of the projects).[2]

The temptation to perform the calculations in Table 3.2 above must be avoided. None of these calculations is correct. Surprisingly, even the one summing the most likely durations is an unreliable estimate of the project completion date, even though that is what scheduling software and manual methods do all the time.

The remainder of this chapter shows an effective and available approach to deriving the most likely, conservative and optimistic duration of the project given the risks embodied in the three-point estimates of duration. It is the Monte Carlo simulation approach that can be found in several software packages.

What is Monte Carlo Simulation?

Monte Carlo simulation calculates the uncertainty in the results of a system where the elements of the system are influenced by numerous uncertain components or elements interacting and operating according to specified rules to make the resulting status of that system uncertain. Monte Carlo simulation is perfect for determining the project completion date since that date is determined by the uncertainty in the durations of many activities that have already been linked logically in the CPM schedule. (See Chapter 8, 'Using risks to drive the analysis and prioritize risks: introducing the risk driver method', to see how Monte Carlo simulation can easily be applied to the risks themselves rather than to the uncertain durations.)

---

2   The statement that the combined extremes occurring is remote depends somewhat on the assumption that these activities' durations are independent of one another. That means that there is not a strong force acting on each of these four activities driving them all in one direction or another, and that none of these being long or short will drive the others. We will explore the concept of correlation between the activity durations in Chapter 10 on correlation in schedule risk.

Monte Carlo simulation was developed in the 1940s but was out of reach of most practitioners until personal computers and associated software became available.[3]

Monte Carlo has been applied to many industrial, scientific, logistical and social systems for years. In recent years Monte Carlo simulation has been applied to project management uncertainty issues such as the project completion date and the project cost estimate.

At its base, Monte Carlo simulation is a brute force method of determining the uncertainty of the project completion date (or cost estimate). Fundamentally, a Monte Carlo simulation simply computes the CPM schedule many times, using all possible combinations of the uncertain activity durations, and records the results for completion dates in simple charts or tables indicating the frequency that different results (for example, completion or milestone dates) occurred.

Applied to project schedules, this means that the schedule is run or iterated (a simulation is made up of many, usually thousands of iterations) many times using different durations as inputs each time. The durations are taken from the probability distributions, such as those in Table 3.1, which are derived from the risk interviews. Each time the schedule is calculated different completion and milestone dates may occur because of the different inputs. A 'histogram' or probability distribution of these different dates represents the possible results that are driven by the risky inputs.

The basic process of Monte Carlo simulation is simple, brute force repetition so it should be easy to explain to management or a customer. To be a little more formal about the process, there are five simple steps in a Monte Carlo simulation:

1. Represent the uncertain components of the schedule, the activity durations, as probability distributions. (As mentioned above, Chapter 8 will apply this method not to the activity durations but to the identified risks themselves.) Different distributions may be used, as dictated by the risk interviews. These are usually constructed from three-point estimates such as those discussed in Chapter 2.
2. For the first iteration, select values of durations at random from the probability distributions for the activities with uncertain durations. (Obviously this method requires specialized simulation software. Several such programs have been developed recently that are compatible with popular scheduling packages.) These durations may be the most likely durations in the schedule but typically they will be different values.
3. Compute the schedule as if it were a standard CPM exercise with exact duration estimates using the durations chosen in Step 2, calculating the schedule dates, phase durations and critical path consistent with those durations.
4. Repeat Steps 2 and 3 many—usually thousands of—times, selecting durations at random and computing a CPM solution for the project each time. There will be many results. The dates or durations can be shown as probability distributions. Each iteration has, potentially, a different critical path as well and these are recorded.
5. Make a histogram or probability distribution of the results of the schedule computation

---

3   'Any method which solves a problem by generating suitable random numbers and observing that fraction of the numbers obeying some property or properties. The method is useful for obtaining numerical solutions to problems which are too complicated to solve analytically. It was named by S. Ulam, who in 1946 became the first mathematician to dignify this approach with a name, in honor of a relative having a propensity to gamble. Nicolas Metropolis also made important contributions to the development of such methods.' Weisstein, E. (2008). Monte Carlo Method, Wolfram MathWorld.

for each of the iterations. At the end of the simulation the results are summarized in charts and tables and analyzed.

Note: The simulation program will do this duration selection, calculation and summation/reporting process.

The results of a simulation provide realistic and important insights under many combinations or scenarios related to the risk of the inputs:

- dates for milestones including the finish milestone;
- durations of various spans of activities or project phases;
- identity of the critical path;
- further analysis of the results can provide statistical results such as the correlation between activity durations and important dates such as the total project completion date. This is useful for understanding the sensitivity of the results to particularly risky activities on particularly critical paths.[4]

Example of a Monte Carlo Simulation

Each run or calculation of the schedule is an 'iteration' and the total of all iterations is the 'simulation.' An illustration of a few iterations from a simulation will clarify the process. Take the simple four-activity schedule above. Since the activities are logically linked finish-to-start the duration of the project in each iteration is simply the sum of their durations, randomly selected for that iteration. Since we do not know which scenario will actually occur, we select many different combinations of inputs—these are called 'iterations.' The first seven iterations from a simulation are shown in Table 3.3.[5]

So far, in seven iterations, no Total Duration has been calculated more than once although some individual activity durations have been selected twice, for example, see Design at 36 days or Build at 50 days. If additional accuracy, for example, minutes rather

Table 3.3   Example iterations from the simple four-activity schedule

Activity	Iterations						
	1	2	3	4	5	6	7
Design	42	37	33	35	27	36	36
Build	50	55	50	46	51	45	44
Test	28	33	28	36	21	39	21
Deliver	12	11	9	11	10	13	10
TOTAL DURATION	**131**	**136**	**120**	**128**	**108**	**133**	**112**

---

4   The 'critical path' is a concept in CPM scheduling that identifies the path(s) on which an extension or elongation will delay the project completion. Different definitions, such as the longest path or the path with the least float, indicate a lack of flexibility on the critical path.

5   These results were generated by Crystal Ball®, a product of Decisioneering, Inc. recently acquired by Oracle, which simulates Microsoft Excel® spreadsheets.

than days, were specified it would also be clear that no input duration has been selected more than once either.

Results of the Monte Carlo Simulation

The important aspect of the simulation, of course, is the result. In the case of the simple four-activity schedule that finishes on September 23 in the static CPM schedule, what do the results of the schedule risk analysis data gathering and simulation tell us?

Figure 3.2 shows the results of a simulation applied to the simple sample schedule.[6]

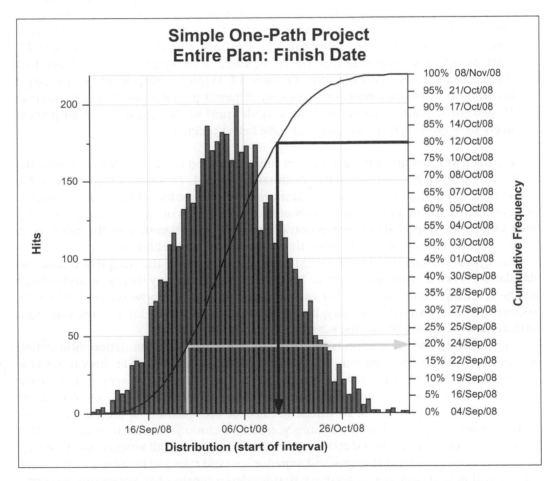

Figure 3.2   Result of simulating the simple 4-activity one path schedule

6   This simulation was conducted using Primavera Pertmaster®. Developed in the UK, Pertmaster is now a Primavera Systems, Inc. product. Primavera has recently been acquired by Oracle.

The results for the simulation show the following picture of this project's schedule:

- The September 23 date calculated in CPM scheduling is not very likely to occur, given the risks that have been specified. From these simulations there is only about a 20 percent likelihood that this project will finish on September 23 or earlier. The probability of schedule success on this project without risk mitigation is problematic.
- The date of September 23 is not even the most likely completion date for this project. From the histograms it seems that the most likely date (the tallest column in Figure 3.2) is closer to October 4 than to September 23. This finding reinforces the statement made above that it is unwise to put confidence in the sum of even the most likely durations as a predictor of the duration of the project.
- Suppose the project manager or customer would like to be pretty sure of completing this project on time. If the plan is set, what date should be adopted as the contract or promise date? A common value for a conservative target is a date that has an 80 percent probability of success, often abbreviated as 'P-80' and that date is about October 12. That means, given the risks that we know, there is an 80 percent probability that this project will finish on or before October 12. While an 80 percent certain target is common, some customers will focus on different probabilities of success such as 50 percent (equally likely to overrun as under-run) or 70 percent (only 30 percent likely to overrun), given the risks that have been specified.

During a Monte Carlo simulation, each time a duration is selected (Step 2 above) the probability distribution for that activity's duration is used. This means that values close to the most likely duration will be picked more often than values close to the optimistic or pessimistic extreme values. The shapes of the distributions of duration uncertainty for each activity are respected in the selection of inputs for each iteration, so the values with more likelihood are chosen more often than those with less likelihood.

Each iteration uses the standard CPM analysis to determine the completion dates, so simulation is not a mysterious process. Simulation just uses many iterations and collects all the resulting data from each. For each run of the schedule those selected values are assumed to be known with certainty, but of course since we do not know those durations with certainty we have to run the schedule many times.

Repetition of the calculation is the strength of Monte Carlo simulation. Simulation is very much like public opinion polling. If you want to know how an upcoming vote for political office will turn out, it is useful to ask many potential voters about their intentions. If you ask only one person how they will vote, the answer will not be a reliable indication of the population of all voters. To improve the reliability of the survey you ask many prospective voters. The more voters you ask the closer you come to approximating the outcome of the election. Of course, you do not want to ask all voters (take a 'census' of voters), both because of the time and expense it would take and because a well-chosen sample will provide a degree of accuracy that renders a census of all voters unnecessary.

When election day comes the results may differ from the results of even the best survey—that is because of randomness and that people can change their minds or fool the survey-taker. That is why we have the actual election (and why they play the entire sports game, no matter what the odds).

Similarly, project results may differ from the results derived from a Monte Carlo simulation no matter how many iterations are performed. It is useful to be somewhat modest in the claims of accuracy for simulation methods while being quite confident that recognizing project risk is much more accurate than ignoring the impact of risk.

Some Mechanics of Monte Carlo Simulation

In practice, the simulation software selects many durations in anticipation of running the model many times. In this discussion we will simplify to only one activity duration and only one iteration.

Selecting the Input Durations for Each Iteration

The process of selecting durations at random for each input and iteration can be illustrated using the first activity, Design. The possible durations that Design can take, given the risk in designing the item to be built, are represented by a three-point estimate with optimistic = 20 days, most likely = 30 days and pessimistic = 45 days. The analyst has assumed a triangular distribution for this activity. The pictures of the Design activity's frequency distribution and its resulting cumulative distribution are shown in Figures 3.3 and 3.4.[7]

To determine the duration that will be used in an iteration for the Design activity, the simulation software mechanically turns the specified input distribution, such as that shown in Figure 3.3, into a cumulative distribution such as that in Figure 3.4. For any iteration the software selects randomly a number between 0 and 1.0 using the

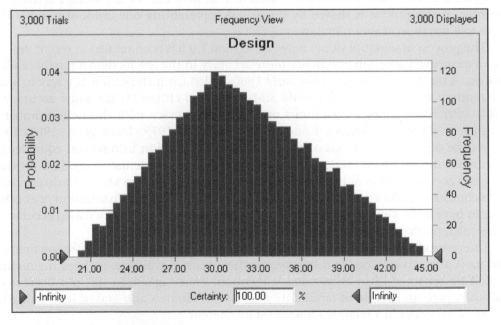

Figure 3.3 Input distribution

---

7   These charts use Crystal Ball® from Decisioneering, Inc., an Oracle company.

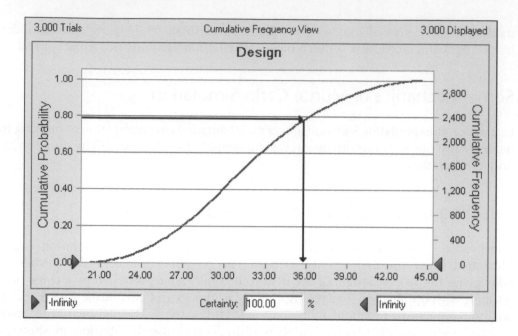

Figure 3.4   Cumulative distribution

computer's random number generator. That value is used to find a value on the Y-axis of Figure 3.4. The cumulative distribution translates that value into a duration on the X-axis of Figure 3.4. The selected duration is used in that iteration for the Design activity. An example of this process is shown by the arrows representing one random number that leads to a duration of about 36 days.

The process of selecting values between 0.0 and 1.0 has been refined in recent years to give a more even selection and hence more accuracy in the results than a purely random process. A method developed in the early 1980s called Latin Hypercube is a way to select the random numbers more efficiently and to develop either: (1) the same accuracy of result with fewer samples (iterations) or (2) better accuracy with the same number of samples or iterations. (McKay, Conover and Beckman 1979; Iman et al. 1981) Latin hypercube organizes the Y-axis of the cumulative distribution into several equal-length partitions and samples at random from each partition in a systematic way. This provides the extra accuracy and avoids the possibility of the purely random Monte Carlo approach from choosing random numbers that clump-up in one area of the distribution or another. If Latin hypercube sampling is available, its use is recommended for increased accuracy of the results, particularly in the tails of the distribution of results.

Interestingly, since Monte Carlo refers to the purely random sample selection method, and most analysts will use Latin hypercube whenever it is available, perhaps the method should be renamed Latin Hypercube Simulation. That suggestion has not been made seriously, perhaps because of familiarity of the term Monte Carlo and of its evoking the image of gambling in a famous Monaco establishment.

How Many Iterations are Needed?

A small number of iterations will not provide much accuracy in the results. For instance, the schedule shown above completes on September 23 if everything goes 'according to plan.' We need to know how likely that is to occur, given the risks in each of the four activities, Design-Build-Test-Deliver. Using a schedule risk simulation program, Risk+[8] on the original schedule we can see the results of a small number, in this case 25, of iterations as follows.

Notice the results. We can actually see each of the 25 iterations, since the height of the bars represent 1, 2, 3 or 4 'hits.' (Notice that 'Each bar represents 2 d' in the upper right of the chart.) The table at the right of this chart shows the dates consistent with the cumulative distribution that snakes through the chart from lower-left to upper-right. The dates in the table indicate that the project is between 20 percent and 25 percent likely to finish on September 23 or before. Is that estimate going to be accurate enough for us? Probably not, since it is like trying to predict a national election by asking 25 potential voters.

How many iterations are needed to develop the required accuracy of the result? This is a difficult question to answer. Mostly, people will compute more iterations for a final report than for interim runs, but that observation still begs the question of how many runs are needed. One measure of the accuracy in the simulation is the '95 percent confidence interval' which is 4.22 days in Figure 3.5. This measure indicates the accuracy in statistical terms of the calculated mean or weighted average date, which is shown as October 5 (see '10/5' the middle of the X-axis of Figure 3.5) to represent the mean of the population of all projects from which these few (25) samples were selected. Using only 25 iterations, the value of the mean completion date of the population of total projects that are as risky as this one is 95 percent likely to be within 4.22 days of October 15, from approximately October 11 to October 19.

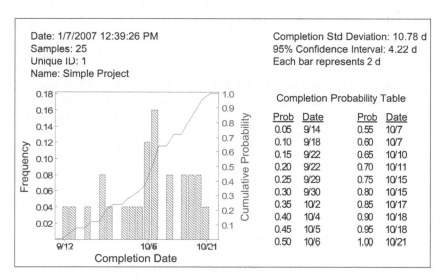

Figure 3.5   Simple four-activity schedule simulated with only 25 iterations

---

8   This simulation was computed using Risk+®, originally developed by C/S Solutions, Inc. which was acquired by Deltek, Inc.

As a partial answer to this question, consider a much larger sample of 5000 iterations on the same project schedule. The results are shown in Figure 3.6. The probability distribution looks a lot smoother and gives more confidence than that with only 25 iterations. The 95 percent confidence interval is now only .28 days, although this more accurate estimate of the average completion date is now October 4, not October 5.

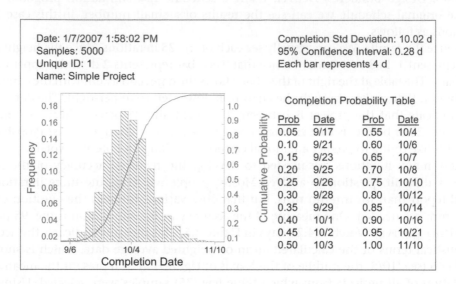

Date: 1/7/2007 1:58:02 PM
Samples: 5000
Unique ID: 1
Name: Simple Project

Completion Std Deviation: 10.02 d
95% Confidence Interval: 0.28 d
Each bar represents 4 d

Completion Probability Table

Prob	Date	Prob	Date
0.05	9/17	0.55	10/4
0.10	9/21	0.60	10/6
0.15	9/23	0.65	10/7
0.20	9/25	0.70	10/8
0.25	9/26	0.75	10/10
0.30	9/28	0.80	10/12
0.35	9/29	0.85	10/14
0.40	10/1	0.90	10/16
0.45	10/2	0.95	10/21
0.50	10/3	1.00	11/10

Figure 3.6   Simple four-activity schedule simulated with 5000 iterations

The results are compared in Table 3.4. Surprisingly, the results may not strike one as being very different from one another.

- A significant difference between these two simulations is in the 95 percent confidence measure. With 5000 iterations we have a 95 percent confidence that our mean is no more than about a quarter of one day from the average day of the population of all possible projects with the same activities, logic and duration risk as our own.
- The average day is different, although is only one day different with 5000 iterations from the result with 25 iterations in this case.
- The probability of finishing on or before the schedule date of September 23 is more accurately calibrated at 15 percent rather than the 20–25 percent range of the smaller sample.
- The 80th percentile has advanced by 3 days from October 15 to October 12 with the 5000 iteration simulation. In other words, if we want accuracy in the tails of the distribution, we need to conduct more iterations.

Review of the two simulations shows that the main message is the same, that September 23 is not a very confident prediction of when this project will finish, and we need until sometime in mid-October to be substantially more confident (80 percent likely) that the project will be done.

Table 3.4   Compare the accuracy of simulations of 25 and 5000 iterations

Comparison of the Results		
	Iterations	
Measure	25	5000
Pr(on or before 9/23)	20%–25%	15%
Mean	5-Oct	4-Oct
80th percentile	15-Oct	12-Oct
Standard Deviation	10.78 d	10.02 d
95% confidence	4.22 d	0.28 d

The number of iterations needed will be the user's discretion, but factors that influence that decision could include:

- the accuracy of the input data, which includes the knowledge of the interviewees and any possible bias in the information gathering, the degree of newness or innovation in the project compared to projects in the interviewees' experience and even the stage of the project's progress since early on there will be a lack of accurate data on risk;
- the degree of accuracy needed in the results, which depends on the rewards or penalties attached to making or overrunning the schedule date;
- the complexity of the project schedule, which may include such modeling tools as probabilistic branching and conditional branching treated in later chapters;
- the amount of time each simulation takes, which depends on the number of activities, the software and the speed of the computer. Happily, today's software specializing in simulation is optimized for speed and today's computers, including laptop versions, run simulation software quite fast.

As a practical matter some 3000 iterations should be enough to handle most schedule risk analysis problems. If there are probabilistic or conditional branches more iterations are needed. Sometimes for final reports more iterations would be preferable if only to give the people receiving the results more confidence since the results rarely change significantly after, say, 1500 iterations.

Measures of Extreme Optimism and Extreme Pessimism

Common measures of extreme optimism and pessimism use the 5th percentile and the 95th percentile results. For this simple schedule the results stated in terms of days' duration are shown in Figure 3.7.

From the simulation, we have a recommended measure of the overall risk of a project's schedule. Certainly 5 percent is a very optimistic standard, because only 5 percent of the projects with these risks will finish in 108 days or earlier. And 95 percent is a very conservative standard, since only 5 percent of the projects will finish in more than 143 days with these risks. (Some organizations use the 10th percentile and the 90th percentile to show extremes, and these are 111 days and 139 days.) There is essentially a 1-month difference between these two extreme possible project durations for this not-very-risky one-path four-activity schedule.

Why are the recommended extreme values of the project path that are derived from the simulation so much narrower than the sum of the optimistic durations and of the pessimistic durations shown—but NOT RECOMMENDED—in Figure 3.2? Those durations were 85 days and 175 days, or about 3 months difference, as shown above in Table 3.2.

The extreme optimistic and pessimistic durations, from P-5 of 108 days to P-95 of 143 days, reflect a characteristic of Monte Carlo simulation and of real projects as well. The activity durations will vary from project to project. However, there is a good amount of cancelling-out of durations that are long within their distributions by those that are short within theirs as the activities are completed along a path. If one activity takes a long time as defined within its own probability distribution the impact of that fact on the project completion date may be cancelled out by some other activity taking short or moderate durations within its probability distribution.

It is unlikely for all activities' durations to be highly correlated with each other so that whenever one is in the upper (lower) end of its distribution the other activities' durations will always be at the upper (lower) end of their own distributions together. If activities'

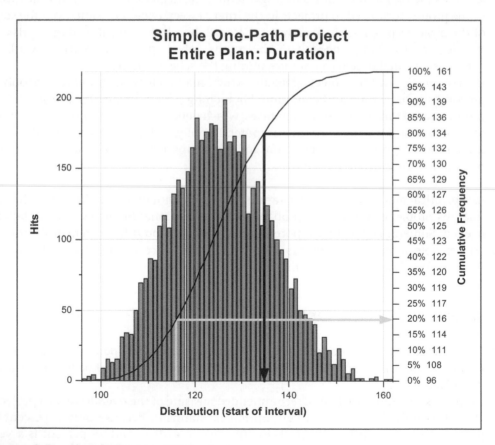

Figure 3.7   Simulation results in days

durations were correlated in this way, the simulation extremes would be wider.[9] Reporting the sum of columns, 85 days on the optimistic end and 175 days on the pessimistic end, would be irresponsible and unwisely cloud stakeholders' judgment about the project. The range from 108 days to 143 days is sufficient to encompass 90 percent of the variation that could be expected in this project.

Usually the range from P-5 to P-95 is a responsible measure of overall schedule risk for a project. We should remember to 'never say never' on the pessimistic end, since some activities may take a year to complete or may never actually complete and the project may fail. This is a recognition of The Black Swan phenomenon (Taleb 2007).

Summary of Monte Carlo Simulation Applied to a Simple One-path Project

Project risk analysis has the power to answer questions that traditional CPM scheduling cannot address well:

* What is the likelihood that we will meet our target schedule? CPM scheduling gives one target date only.[10] Monte Carlo simulation gives a good estimate of the probability that the project will finish on any date or earlier/later, based on the risks uncovered in the interviews and used in the analysis. Interviewing is discussed in Chapter 5.
* What contingency in time should be put on the CPM completion date to provide the degree of confidence in the date that we require? CPM scheduling provides one completion date only. Any contingency reserve based on that date is a guess at best. Monte Carlo simulation provides a menu of possible dates with the probability that the project will finish on those dates or earlier/later. It is up to the decision makers to determine the level of certainty that is required, but some people suggest adopting the 80th percentile. A schedule contingency reserve activity could be added at the end of the schedule representing the amount of time needed to provide for the risk in the schedule.

Usually project managers do not welcome contingency reserves of time in their schedules with nearly the enthusiasm that they welcome (require) that reserves be put on their project's cost estimate. Perhaps using these techniques the concept of a contingency of time will be more widely accepted by project managers and stakeholders.

Project schedules are uncertain, in large part because the activity durations are estimated but cannot be assured. Since durations are uncertain, inquiries made in interviews or workshops about the degree of uncertainty can provide some measures of optimistic, most likely and pessimistic (low, most likely and high) durations for each uncertain activity. The risk analyst turns to specialized software that is available to simulate schedule risk. The risk analyst specifies the probability distribution represented by the three-point estimates and the simulation software does the rest.

---

9   Monte Carlo simulation models can be configured to reflect correlation of the durations between pairs of activity, and a later chapter describes the correlation capability. Correlation is discussed in Chapter 10.

10   Review of path float can be indicative of schedule pressure, but it results from the structure of the schedule and does not provide for project risk as, for example, the P-80 result from a Monte Carlo simulation would do.

Monte Carlo simulation is a well-established and best-practice standard method of analysis that develops schedule risk results by examining the implications of many, usually several thousands, of combinations of activity durations for risky activities. While the actual implementation in software is sophisticated, the method is brute force. The results are provided by the software, leaving the analyst the tasks of collecting the data and interpreting and communicating the results.

Often the schedule risk analysis shows that the project team may have some difficulty in accomplishing the schedule in the time indicated by CPM scheduling. In a simple single-path schedule used in this chapter the results were illuminating. The schedule reviewed in this chapter contained only four activities and simple, finish-to-start logic. Each activity had duration risk that is modest given some of the risk found in actual schedules. The durations given in the interviews confirmed the most likely values found in the CPM schedule. Yet:

- the CPM date of September 23 was not even the most likely date—that was about October 4, about 11 days later;
- the CPM date had only about a 15 percent chance of occurring given the present plan and known risks;
- a contingency allowance up to October 12 would be needed to provide a comfort level of 80 percent probability of success.

It is clear that relying on CPM scheduling for accurate results in making commitments is a risky venture. Schedule risk analysis provides much more information that is useful to the project manager and project stakeholders.

Schedule risk analysis does require collecting more data and mastering specialized software, but the value in terms of providing realistic answers to important questions is very clear. Any person realizing this fact should perform schedule risk analysis on any project for which finishing on time is important.

In Chapter 6, risk in a more realistic schedule with parallel paths is discussed. At schedule merge points the risk is generally greater than it is along single paths. Hence, single-path risk is just the beginning of the story of schedule risk.

References

Iman, R. L., et al. (1981). 'An Approach to Sensitivity Analysis of Computer Models, Part I. Introduction, Input Variable Selection and Preliminary Variable Assessment.' *Journal of Quality Technology* 13(3): 174–183.

McKay, M. D., et al. (1979). 'A Comparison of Three Methods for Selecting Values of Input Variables in the Analysis of Output from a Computer Code.' *Technometrics* 21: 239–245.

Taleb, N. N. (2007). *The Black Swan: The Impact of the Highly Improbable*. New York, Random House.

Weisstein, E. (2008). Monte Carlo Method. Wolfram MathWorld.

4 *A Good Project Schedule is Needed: Critical Path Method Scheduling 101*

Schedule Risk Analysis is Based on a Good Project Schedule

A project schedule is a dynamic representation in time (forward-looking) of the project's execution plan, including all activities required to complete the project, logical relationships between those activities, estimates of activity duration, resources required to perform activities, calendars, and so on. A project schedule is not simply a series of dates on which activities will be completed, but the system of information that produces these dates, among other important outputs.[1]

Project scheduling is a required tool of project management. Every project manager, sponsor, team leader and customer agrees that a project needs a schedule. Most schedules these days are developed and maintained in project scheduling software.[2] Most large projects employ several project schedulers. Small projects have a partial schedule of some sort. The schedule predicts the dates of completion and other milestones for the project. It is also used to manage daily activities, to plan resources and to record status. When the schedule shows that an important milestone is not likely to be met, project managers may use the schedule to examine alternative approaches in a 'what-if' exercise. The schedule agreed to by the contractor and customer is called the baseline schedule.

Schedules also provide an update based on events by frequent statusing with actual progress on activities that are complete or in-process. Statused schedules, generally called the 'current schedule,' may imply different completion dates from the baseline schedules. Variances of the current schedule from the baseline are important for analysis, prediction and mid-course correction. Ultimately, when the project is completed, the schedule variances can form the basis for resolution of liquidated damages resolving the issues between the parties as to which was responsible for a late completion.

Unfortunately, many schedules do not completely follow the rules of project scheduling and are unusable, or more properly are not yet ready for use, as critical path method (CPM) schedules or in a schedule risk analysis. The rules of scheduling are

---

1　The PMBOK® Guide of PMI defines project schedule as: 'The planned dates for performing schedule activities and the planned dates for meeting schedule milestones.' This is inconsistent with common parlance, however, and most professional schedulers use the term 'project schedule' in the sense indicated in this section. PMI (2004). *A Guide to the Project Management Body of Knowledge*. Newtown Square, PA, Project Management Institute.

2　Although it is marginally possible to schedule using manual methods, those are out of date and have been superseded by scheduling software from the easy and accessible to the sophisticated and more difficult. We will assume that, as a practical matter, all projects use automated methods.

expressed in a number of documents and schedule assessment software readily available to the scheduling community.

A simple test of the readiness of the project schedule is whether it predicts the correct dates and critical path when activity durations change. Schedules that do this are properly called 'dynamic schedules' since they continue to be accurate as a dynamic project progresses and durations change from their baseline estimates.

In contrast to a dynamic schedule, many project's schedules are more accurately characterized as pictures, generated in scheduling software, showing activities finishing on important project dates. Schedulers often use constraints such as 'must finish on' a certain date or 'finish not later than' another date to make the picture agree with expectations or contracts. When durations change, these project schedules do not produce the correct results without manual intervention by the scheduler. In fact, some schedulers are under instruction to keep key milestone dates constant as the predecessor durations lengthen and activities delay. This can be done by a number of mostly unprofessional devices such as shortening the durations of some activities or breaking the logic between activities. Such schedules result in ineffective and unprofessional CPM scheduling. They are not really schedules and certainly are not ready for schedule risk analysis.

Experience shows that most schedules do not have the tight logical structure that is required for both CPM scheduling and for risk analysis using Monte Carlo simulations. One of the most common faults is the presence of 'dangling activities' where the logical relationships are incomplete compared to international standards. In this chapter we see some of the problems commonly encountered when the schedule risk analyst first meets the project schedule and how the schedule logic can be improved for application of Monte Carlo simulation.

CPM Project Scheduling

WHAT IS REQUIRED FOR A PROPER PROJECT SCHEDULE?

Precedence diagramming and CPM scheduling is the standard today. It has been an established process of project management for a long time. Scheduling is perhaps the most common technique of project planning and control and is practiced by thousands of individuals with more or less training, experience and expertise in the discipline. Scheduling software is readily available and the companies that make the software are more than happy to train schedulers in their newest versions.

With this rich experience and access to excellent scheduling tools and with so much riding on successfully completing the project on time, why are so many schedules essentially cartoons, basically illustrations of dates that seem to support project objectives? Why are they not tools of analysis and forecasting that respond correctly to changed data to produce the correct dates and critical paths?

Scheduling problems arise because:

- Project scheduling is a difficult discipline. The scheduler must keep myriad activities, resources, logical relationships and constraints consistently and completely stated in one software system. Even an expert scheduler is challenged with managing a large, detailed schedule.

- Data about schedules come from many sources, often from teams or subcontractors who do not understand scheduling or are not able or willing to give the data to the scheduler. The scheduler must be an expert communicator and persuader of people for whom the schedule is not their highest priority.
- Management or the customers often insist on unrealistic deadlines. Faced with such institutional pressure, schedulers will often put together a schedule with 'magic numbers' that has little possibility of success and little credibility among the project teams, although it shows the 'right dates' as dictated by management.
- The discipline of proper project scheduling logic is demanding and precise. It is not always well-understood by practitioners. Schedulers are often self-trained or are trained by others who do not fully understand the discipline. Those with good training may have forgotten the requirements of good schedule logic or do not use it in the heat of trying to get a schedule done by a deadline.

These points are discussed in turn.

SCHEDULING IS A DIFFICULT DISCIPLINE

Schedulers must be masters of the scheduling discipline. It is a challenge to put together a large detailed schedule and to maintain it accurately during the action and pressure of project planning or execution. Often, the successful scheduler knows more about the workings of the project than does the project manager because scheduling touches every aspect of the project at a level of detail not necessary to overall management.

Gaining the detailed information needed for accurate scheduling implies the need to understand all of the processes and activities of the project and to be able to get information about these from the people involved. Some schedulers come from the disciplines active in the project, such as software or process engineering, and therefore understand the nature of the work. Some schedulers have been involved with such projects before, and are able to draw on their own experience to develop a schedule that includes the activities typically required in such a project.

Sometimes, on the one hand, a person who is not familiar with the type of project is assigned to the scheduling function. With little experience from which to draw, this person has to learn many disciplines quickly. On the other hand, the scheduling role may be assigned to a person expert in the processes of the project but not familiar with, or interested in, the scheduling art. Either of these two situations can lead to poor scheduling.

Project scheduling is not necessarily well-supported by the project organization. Some schedulers have plenty of help. They can take time to sit with discipline leads in order to learn from them how they want to do the project. Some other schedulers may be thrown into the fray without much support. Sometimes schedulers are assigned to multiple projects and need to keep each project team happy—this situation often leads to throwing together a schedule in haste, patching together schedules of earlier projects, with little time for understanding the specific project or for application of proper scheduling discipline. Scheduling is sometimes viewed by management as an essential but not necessarily valuable skill.

Schedulers differ widely in their abilities. Some project individual may be presented with a scheduling software package and told to become a scheduler overnight.

Some people finding themselves in scheduling jobs would rather be assigned elsewhere on the project. Some do not have the temperament of care and precision necessary to be successful. Others do not have the people skills that are absolutely required to get along with project team leads and project managers. Often schedulers are not trained or are poorly trained. Even schedulers who go through rigorous scheduling courses may not understand what they need to do or how important it is. A good scheduler is both a rare resource and extremely important to the project's success.

SCHEDULE DATA COMES FROM MULTIPLE SOURCES

Schedulers must be good at interacting and communicating with people. Project team leaders and members have data the scheduler needs to complete a part of the schedule. Such data includes the activities to be completed, the sequence in which they are completed, their duration in working days, resources required, whether they obey different calendars from the rest of the project and any constraints that may be important. During execution the scheduler needs to status the schedule, collecting data on actual events and any changes in schedule logic.

A good project scheduler must be able to get project-related data from project participants such as team leads who are usually busy and often do not have time or the inclination to support the schedule. This attitude toward the scheduling function is not cavalier; it may be driven by the necessity of progressing the project.

The scheduler must communicate with these busy people in ways that elicit the correct information in a timely way with the least burden on the participants. The scheduler must have the team leads review the schedule and give it their blessing. These interpersonal transactions are many and constant throughout the project, both in the planning and execution phases. A successful scheduler has a good way with people, resulting in team leads who want to participate in this aspect of project management. People who are good technically with scheduling software are not necessarily good schedulers.

MANAGEMENT MAY INSIST ON UNREALISTIC SCHEDULES

Discussions with project schedulers often turn to those projects on which management, the customer or other stakeholders insist on completion or milestone dates that are not realistic under any circumstances. Such projects usually result in pressure on the scheduler to make a schedule that is not believed by anyone. Activities' durations are shortened because 'that's all the time we have.' Activities and paths are overlapped, for instance, making the construction start before the design is completed, a practice called 'fast tracking.' If fast tracking works it can speed up the project but problems, such as re-work, occur when it does not work.

There is a real tension between project stakeholders and schedulers when the former hold unrealistic expectations. The organization wants the project to finish on or before a date that is determined by some necessity such as commercial needs, strategy needs, regulatory need or just plain management style. Some deadlines are literally determined by the alignment of the planets—scientific exploration of planets depends on making a launch of a payload when the planets are in a certain relationship in their orbits, often leading to a very narrow window for successful delivery of a payload. This is real pressure on the project schedule.[3]

Project scheduling is a quasi-profession. That means that schedulers are ultimately serving the needs of the project manager. In this relationship it is difficult to resist the pressure to produce a schedule that supports management's sometimes unrealistic dates. There is no particular code of conduct that the scheduler must adhere to that provides support to a scheduler if they must say 'no' when management says: 'That's all the time we have, we have no option.' The scheduler may use powers of persuasion, backed up with history and data, but ultimately it seems that the project manager determines the dates of the initial or baseline schedule. Unreasonable dates can be established, and may even be maintained for some time during execution, in the face of facts to the contrary. In this situation schedulers are expected to follow orders even if that means producing schedules that have no chance of success.

Some, but perhaps few, schedulers have the option and the fortitude to resign in the face of improper management pressure. Contrast the situation of a project scheduler with that of a structural engineer, a licensed professional engineer (P.E.), who may be asked to certify an unsafe design. That P.E. will lose the valuable license to practice by signing an unsafe design. The P.E. just will not do it, and the unsafe design cannot be built. The scheduler does not have the same leverage against unsafe schedules.

One resolution of the tension between scheduler and stakeholder is the crafting of assumptions that would have to be true if the schedule were to be short enough to satisfy the stakeholder. The scheduler knows that the assumptions (for example, 'We get the compressor early,' 'Resources are double what we have right now,' or 'We don't really need that permit.') are at least questionable if not actually disprovable. A good schedule would require management to sign on to these assumptions by taking actions that validate the assumptions.

The structure of a conditional statement, 'if, then, else' is a game that the scheduler can play to propose the assumptions that would be required to meet the unrealistic demands of management for a short schedule. The problem arises when management sees the schedule and believes that the assumptions underlying it must be true or they would not have been assumed. The assumptions become promises of those who made them, rather than conditional assumptions. This asymmetrical treatment of assumptions (the scheduler makes a conditional statement but management believes the statement is a promise) makes for a short-term truce but long-term trouble as the assumptions turn out to be false and the recriminations start. The scheduler is in a weak and vulnerable position when management comes calling with the accusations because project scheduling is not yet fully a profession with legal requirements, responsibilities and protections.

PROJECT SCHEDULING DISCIPLINE AND ABUSES

Project scheduling is a discipline with rules that make the schedule a tool of forecasting and analysis. The discipline is understood, taught and practiced well by many people. Unfortunately there are abuses in scheduling that are all too common. If present experience

---

3   Projects with immovable deadlines have the unpleasant choice of adding more resources to finish the job or de-scoping the project with the result that the finished product has less functionality (scope) or quality than originally planned, or both.

is a guide, most project schedules exhibit a degree of scheduling abuse that makes the forecasts of project dates and identification of the critical path unreliable. Many, if not most, schedules cannot be reliably simulated in a schedule risk analysis without at least some corrective actions on their logic.

There are several problems with practical project scheduling that are typically found and must be corrected to have both a good CPM schedule and a reliable schedule risk analysis.[4] This chapter illustrates some of the problems and extends them to the impact on project schedule risk analysis that uses Monte Carlo simulation (see Chapter 3).

Overuse of Late-Date Constraints

Scheduling software provides schedulers with constraints of several types. Often there are set dates when important events must happen. These dates are usually on important milestones such as project design reviews or decision points, deliveries and of course the project completion date. There are implications to these dates' overrunning, and management does 'not want to see' the dates overrun in the schedule. It is tempting, and the scheduler is sometimes under instruction, to set a constraint in the schedule that the event will not, at least on the computer screen, finish later than that date. Finishing earlier would be acceptable, but there is often not much chance of that happening.

One way to schedule to meet the contract's 'need dates' is to put a constraint on the activity such that it 'finishes not later than' or 'must finish on' the date in the contract. Schedulers often put these constraints on the finish dates of major project events during schedule development. These two particular constraints affect the 'late dates' or the 'backward pass' using the terminology of CPM scheduling. That means that any schedule that overruns that date will show negative float (called slack in some software packages) indicating that the scheduler has yet to find ways to achieve management's or the contract's date. The scheduler must change the schedule logic or make other changes (often shortening durations if more resources could be made available or that the available resources will be more productive) to meet the required date. The scheduler is not finished if there is any negative float, so late-date constraints are useful devices to employ during schedule development.

If the need dates are feasible, the finished schedule that results from resolving negative float supports the key dates. After the scheduler is finished, the constraints can be taken out and the 'correct' dates will be calculated without being constrained.

Supposing that the dates are supportable, we often find that schedulers tend to rely on these constraints even after the initial schedule development. These dates constrained in the schedule take on rigidity that they do not have in the real project. A constraint is an artificial device that exists only on the computer program. Placing a constraint in the scheduling software program may make the schedule look good on the screen but it does not keep the actual project from being late.

If the need dates are not feasible with the available resources and work to be done, even after exploring different approaches to the work plan, the scheduler must inform

---

4   This section is not intended to be a primer on project scheduling. There are many aspects of good scheduling that do not appear here.

management that corrective action rather than imaginative scheduling will be needed. Otherwise, 'magic numbers' that nobody really believes would result.

If there is a problem meeting the dates, either in CPM scheduling or in the risk analysis, it is clearly better to show the project finishing late on the scheduler's computer screen rather than to experience slippage in the project itself. Showing a project's schedule problems clearly and realistically in the schedule early in the project offers the possibility of fixing those problems. If the scheduler is forced to use late-date constraints after the schedule baseline has been set, they may communicate to management that the project is late by focusing on the paths with the most negative float. Showing negative-float paths indicates clearly that the project is not finishing on time while technically adhering to a date that has proved to be unrealistic.

Use of late-date constraints has the potential to encourage scheduling abuses. For example, sometimes the key event moves beyond its constraint date when a scheduler statuses the schedule or reschedules some activities based on new information. Management has been known to insist that the scheduler adjust (shorten) some activity duration on the critical path so the constraint date is not violated. The practice of shortening the schedule arbitrarily (that is, without the promise of added resources like double-shifting or de-scoping the activity) to make the schedule look as if it supports a constraint date is a serious abuse of scheduling. Some strong schedulers will leave the constraint in and report on those paths that have the most negative float, highlighting the problem without technically moving the date.

Leaving in the constraint date also frustrates schedule risk analysis under certain circumstances. Consider the simple schedule in Microsoft Project shown in Figure 4.1.

This project can be left unconstrained with the uncertainty on activity durations shown in Table 4.1 (assuming three-point estimates and triangular distributions are used).

An unconstrained simulation of this schedule shows the following results (Figure 4.2).[5] The average completion date is November 21 and the 80th percentile is December 1.

ID	Task Name	Duration	Start	Finish	May	June	July	August	September	October	November
1	**Project**	**150 d**	**6/1**	**10/28**							
2	Start	0 d	6/1	6/1	6/1						
3	**Unit 1**	**95 d**	**6/1**	**9/3**							
4	Design Unit 1	30 d	6/1	6/30							
5	Build Unit 1	40 d	7/1	8/9							
6	Test Unit 1	25 d	8/10	9/3							
7	**Unit 2**	**95 d**	**6/1**	**9/3**							
8	Design Unit 2	30 d	6/1	6/30							
9	Build Unit 2	40 d	7/1	8/9							
10	Test Unit 2	25 d	8/10	9/3							
11	**Integration and Test**	**55 d**	**9/3**	**10/28**							
12	Begin Integration	0 d	9/3	9/3					9/3		
13	Integrate Units 1 & 2	30 d	9/4	10/3							
14	Test System	25 d	10/4	10/28							
15	Finish	0 d	10/28	10/28							10/28

Figure 4.1   Simple two-path one merge point schedule

5   Different Monte Carlo simulation packages treat constraints differently. For instance, Pertmaster overcomes the 'must finish on' and 'finish no later than' constraints on the final milestone. The reader is encouraged to set up a simple schedule such as the one above and try this experiment with the package used to see the effect of late-date constraints on their results.

Table 4.1   Uncertainty ranges on the simple schedule

ID	Task Name	Rept ID	Min Rdur	ML Rdur	Max Rdur	Curve
1	**Project**	2	**0 d**	**0 d**	**0 d**	0
2	Start	0	0 d	0 d	0 d	0
3	**Unit 1**	0	**0 d**	**0 d**	**0 d**	0
4	Design Unit 1	0	20 d	30 d	45 d	2
5	Build Unit 1	0	35 d	40 d	50 d	2
6	Test Unit 1	0	20 d	25 d	50 d	2
7	**Unit 2**	0	**0 d**	**0 d**	**0 d**	0
8	Design Unit 2	0	20 d	30 d	45 d	2
9	Build Unit 2	0	35 d	40 d	50 d	2
10	Test Unit 2	0	20 d	25 d	50 d	2
11	**Integration and Test**	0	**0 d**	**0 d**	**0 d**	0
12	Begin Integration	0	0 d	0 d	0 d	0
13	Integrate Units 1 & 2	0	20 d	30 d	45 d	2
14	Test System	0	20 d	25 d	50 d	2
15	Finish	1	0 d	0 d	0 d	0

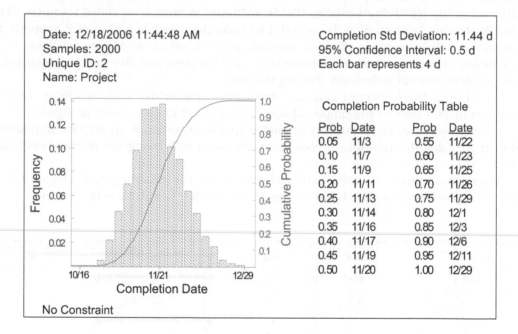

Figure 4.2   Risk analysis of a schedule without constraints

Suppose the project customer placed a 'finish not later than' (FNLT) constraint date of October 28, matching the static schedule's completion date, on the finish milestone. Most schedule risk analysis software packages calculate the early finish dates. In this case, Risk+ simulating Microsoft Project® schedules, a simulation during which results are collected at the project summary bar level will not be affected. This is shown in Figure 4.3, which shows exactly the same results as those for the unconstrained schedule above.

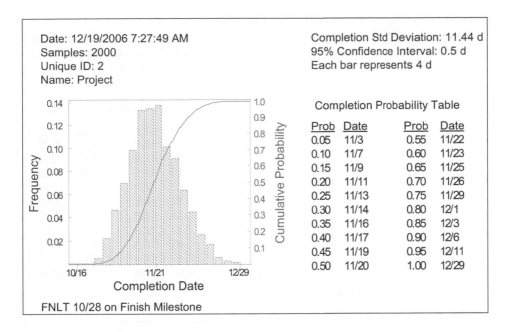

Date: 12/19/2006 7:27:49 AM
Samples: 2000
Unique ID: 2
Name: Project

Completion Std Deviation: 11.44 d
95% Confidence Interval: 0.5 d
Each bar represents 4 d

Completion Probability Table

Prob	Date	Prob	Date
0.05	11/3	0.55	11/22
0.10	11/7	0.60	11/23
0.15	11/9	0.65	11/25
0.20	11/11	0.70	11/26
0.25	11/13	0.75	11/29
0.30	11/14	0.80	12/1
0.35	11/16	0.85	12/3
0.40	11/17	0.90	12/6
0.45	11/19	0.95	12/11
0.50	11/20	1.00	12/29

FNLT 10/28 on Finish Milestone

Figure 4.3   Simulation with FNLT constraint

On the Finish Milestone and Collecting Data on the Summary Task

However, the accuracy of these results depends seriously on where the data are collected. If the data are collected at the finish milestone date, there will be a dramatic effect on the simulation results. The finish milestone is risky and unconstrained will finish later than the constraint date about 95 percent of the time, but the constraint will not let it do so in Microsoft Project®. For the finish milestone, the results of the same simulation are presented in Figure 4.4.

On the Finish Milestone and Collecting Data on the Finish Milestone

The results in Figure 4.4 are dramatic and clearly wrong. But the error of leaving the constraint on the milestone is easily discovered if the constraint is placed on the activity where data will be collected. If the 'finish not later than' or 'must finish on' constraint is placed on some activity in the middle of the schedule, the results for the completion milestone or total project summary task may look to be valid but they will be biased in that they will underestimate the risk of the project as a whole. This is shown by comparing Figure 4.2 with Figure 4.5.

Suppose the constraint is placed on the Begin Integration activity, forcing it to finish on or not later than September 3, the date scheduled for that activity. Notice that the simulation results generally look as expected (a probability distribution with approximately normal shape) and hence they give no indication that there is anything wrong.

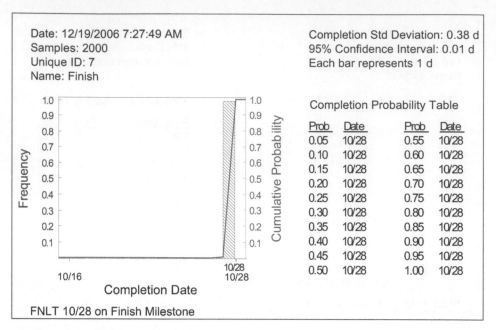

Date: 12/19/2006 7:27:49 AM
Samples: 2000
Unique ID: 7
Name: Finish

Completion Std Deviation: 0.38 d
95% Confidence Interval: 0.01 d
Each bar represents 1 d

Completion Probability Table

Prob	Date	Prob	Date
0.05	10/28	0.55	10/28
0.10	10/28	0.60	10/28
0.15	10/28	0.65	10/28
0.20	10/28	0.70	10/28
0.25	10/28	0.75	10/28
0.30	10/28	0.80	10/28
0.35	10/28	0.85	10/28
0.40	10/28	0.90	10/28
0.45	10/28	0.95	10/28
0.50	10/28	1.00	10/28

FNLT 10/28 on Finish Milestone

Figure 4.4   Simulation with FNLT constraint

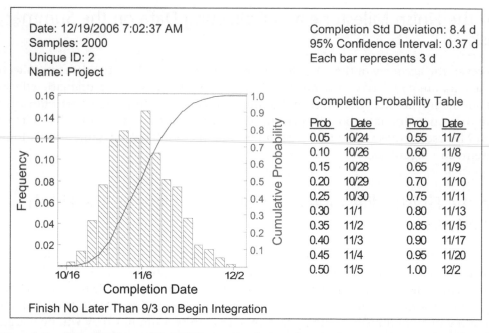

Date: 12/19/2006 7:02:37 AM
Samples: 2000
Unique ID: 2
Name: Project

Completion Std Deviation: 8.4 d
95% Confidence Interval: 0.37 d
Each bar represents 3 d

Completion Probability Table

Prob	Date	Prob	Date
0.05	10/24	0.55	11/7
0.10	10/26	0.60	11/8
0.15	10/28	0.65	11/9
0.20	10/29	0.70	11/10
0.25	10/30	0.75	11/11
0.30	11/1	0.80	11/13
0.35	11/2	0.85	11/15
0.40	11/3	0.90	11/17
0.45	11/4	0.95	11/20
0.50	11/5	1.00	12/2

Finish No Later Than 9/3 on Begin Integration

Figure 4.5   Simulation with FNLT constraint on begin integration and collecting data on project summary task

Upon closer inspection of the constrained schedule simulation results, the average completion date is November 6 and the 80th percentile is November 13, earlier than the unconstrained results of November 21 and December 1 respectively as shown in Figure 4.2.[6] The difference and the optimistic bias is because the constraint in the middle of the schedule forces the successor to the Begin Integration milestone to start earlier, on the day after the constraint (September 4), than they would if they were unconstrained.[7]

In summary, the best approach to simulating a schedule with late-date constraints ('finish no later than,' 'must finish on' or 'mandatory finish') is to delete the constraint before the simulation exercise. Although under some circumstances the simulation software will be correct even with those constraints, the user should check the specific simulation software to be sure if the constraints are to be left in. Late-date constraints, if they have any effect, will produce underestimates of the risk in the schedule.

Dangling Logic

Many schedules are put together so that the activity bars and milestones appear on the 'right' date in the baseline. Often the logic is too weak to provide correct results if activity durations change. Schedules must have correct and complete logical relationships between activities (predecessors and successors) in order to provide the right results in a schedule of a changing, dynamic project and of course in a Monte Carlo simulation. Otherwise they only produce the correct results in the static sense, when all goes 'according to plan.' This deficiency in scheduling is often caused by activities with incomplete logic, often called Danglers.

DANGLING LOGIC—EACH ACTIVITY MUST HAVE A SUCCESSOR (AN INTRODUCTORY TEST)

A well-understood but commonly found example of dangling logic is an activity that has no successor. Suppose the scheduler has forgotten to provide a successor to an activity, even though that activity has a successor in the real project. The scheduling software will automatically connect that activity to the project completion date that may be far in the future and compute a large amount of total float (the term Slack is sometimes used).[8]

An example of dangling logic resulting from no successor is shown in Figure 4.6. Notice that Activity 2, Design Component A, has no successor in the schedule, even though it will be required before Activity 4, Integrate the System, in the project. Notice that the Total Float (Total Slack in Microsoft Project®) shows 70 days. This total float is spurious since it is caused by the absence of the successor for Design Component A.

---

6   In Microsoft Project® the constraints can be searched and eliminated individually. Alternatively, the Tools, Options, Schedule tab has a box 'Tasks will always honor their constraint dates' which can be unchecked to disable the constraints during simulation.

7   Interestingly, Pertmaster simulating Primavera P3 files has exactly the same problem as Risk+ simulating Microsoft Project® files with the interior Mandatory 9/3 constraint, although not with a 'not later than constraint.' Again, the reader is cautioned to check the specific simulation software for its constraint handling.

8   Total float indicates that the activity may grow longer or finish later by the amount of total float days while still not delaying the project's completion.

Activity 2 is a dangling activity. The implication of this incomplete logic becomes clear when we explore the implications of Activity 2 taking longer than 50 days, say 80 days.

The schedule in Figure 4.7 continues to complete on November 15 even though a key activity is longer than before. This is because the longer activity is not hooked into its logical successor, Integrate the System, but it has not yet beyond the finish date. In the project, of course, Integrate the System will be pushed out because both components need to be available to integrate into the system for testing.

Correct CPM scheduling requires that changes in activity durations lead to the correct implications for their successor activities and for the project as a whole without manual intervention. In this case, the link between Design Activity A and Integrate the System is missing, so the schedule does not automatically delay the finish date when the design of Activity A takes longer than anticipated. This is faulty or missing logic and Activity A is a dangler. CPM scheduling and Monte Carlo simulation both require successors.

ID	Task Name	Duration	Start	Finish	Successors	Total Slack	2nd Quarter Apr | May | Jun	3rd Quarter Jul | Aug | Sep	4th Quarter Oct | Nov | Dec	1st Q Jan
1	Start	0 d	6/1	6/1	2,3	0 d	6/1			
2	Design Component A	50 d	6/1	8/9		70 d				
3	Design Component B	50 d	6/1	8/9	4	0 d				
4	Integrate the System	40 d	8/10	10/4	5	0 d				
5	Test System	30 d	10/5	11/15	6	0 d				
6	Finish	0 d	11/15	11/15		0 d			11/15	

Figure 4.6   Example of dangling activities Design Component A has no successor

ID	Task Name	Duration	Start	Finish	Successors	Total Slack	2nd Quarter Apr | May | Jun	3rd Quarter Jul | Aug | Sep	4th Quarter Oct | Nov | Dec	1
1	Start	0 d	6/1	6/1	2,3	0 d	6/1			
2	Design Component A	80 d	6/1	9/20		40 d				
3	Design Component B	50 d	6/1	8/9	4	0 d				
4	Integrate the System	40 d	8/10	10/4	5	0 d				
5	Test System	30 d	10/5	11/15	6	0 d				
6	Finish	0 d	11/15	11/15		0 d			11/15	

Figure 4.7   Adding days to dangling activities may have no impact on the schedule

DANGLING LOGIC—ACTIVITIES WITH INCOMPLETE SUCCESSORS

Isn't having a successor activity enough of a requirement? No. The rule that each activity must have a successor is correct as far as it goes, but it is inadequate for complete project scheduling. Many schedulers stop at this point, assuming that if each activity has a successor the schedule logic is complete.

There are more specific requirements than just the existence of a successor to ensure complete and robust logic. The type of predecessor and successor is important for both CPM scheduling and Monte Carlo simulation (Hulett 2007).

Suppose a predecessor has a start-to-start successor only. The result is much like the activity without a successor that was discussed above. If the predecessor is longer than the schedule duration the successor is unaffected in the schedule, although it will be affected in the project. The situation is shown in Figure 4.8, using Primavera P3.

Activity ID	Activity Description	Orig Dur	Early Start	Early Finish	Successors	2006 (M JUN JUL AUG SEP OCT NOV)
SRA Book Dangler Test						
Plan the Project						
PLN01	Plan the Project	40	01 JUN06	10 JUL06	UNIT101*	Plan the Project
Unit 1						
UNIT101	Build Unit 1	40	11 JUL06	19 AUG06	UNIT201*	Build Unit 1
Unit 2						
UNIT201	Build Unit 2	40	11 JUL06	19 AUG06	INT02*	Build Unit 2
Integration and Test						
INT02	Integrate Units	30	20 AUG06	18 SEP06	INT03*	Integrate Units
INT03	Test System	25	19 SEP06	13 OCT06		Test System

Figure 4.8   Activities with successors may be dangling example with start-to-start successors

Activity ID	Activity Description	Orig Dur	Early Start	Early Finish	Successors	2006 (M JUN JUL AUG SEP OCT NOV)
SRA Book Dangler Test						
Plan the Project						
PLN01	Plan the Project	40	01 JUN06	10 JUL06	UNIT101*	Plan the Project
Unit 1						
UNIT101	Build Unit 1	80	11 JUL06	28 SEP06	UNIT201*	Build Unit 1
Unit 2						
UNIT201	Build Unit 2	40	11 JUL06	19 AUG06	INT02*	Build Unit 2
Integration and Test						
INT02	Integrate Units	30	20 AUG06	18 SEP06	INT03*	Integrate Units
INT03	Test System	25	19 SEP06	13 OCT06		Test System

Figure 4.9   Activities with start-to-start successors may be dangling

Build Unit 1 has a start-to-start successor, Build Unit 2, so it passes the introductory test described above that every activity should have a successor. The bars are aligned and the picture looks like the right schedule. However, as the chart indicates, the logic is start-to-start.

What happens if Build Unit 1, the dangler, takes 80 days instead of the 40 days that are in the plan? We know that integration should be delayed by 40 days but that is not the case with this dangling schedule logic. The activity Integrate Units does not 'see' the change in Build Unit 1 because it depends only on that predecessor's start date, not its finish date.

Build Unit 1 has passed the test that the successor's column is not blank, but it is a dangler. However, the risk in Build Unit 1 is ignored by this logic; Integrate Units starts (and finishes) before Build Unit 1. The introductory test, that each activity needs a successor, is insufficient in the case of start-to-start successors.

A good scheduler may say: 'I know how to fix this. I will use finish-to-finish logic. That way if Build Unit 1 is longer, at least Build Unit 2 will be pulled out.' Figure 4.10 shows the picture before any elongation in any activity.

Again, the activity bars look to be in correct alignment initially. But what if the dangler, in this case Build Unit 2, takes longer than 40 days? What happens to its successors? The situation is shown in Figure 4.11.

In the case of just finish-to-finish logic, if the dangler Build Unit 2 takes longer than anticipated, the scheduling software just starts it sooner.[9] If Build Unit 2 starts on

---

9   This result assumes that there is no other predecessor keeping Build Unit 2 from starting early and that the activity has not actually started.

Activity ID	Activity Description	Orig Dur	Early Start	Early Finish	Successors
SRA Book Dangler Test					
Plan the Project					
PLN01	Plan the Project	40	01 JUN06	10 JUL06	UNIT101*
Unit 1					
UNIT101	Build Unit 1	40	11 JUL06	19 AUG06	UNIT201*
Unit 2					
UNIT201	Build Unit 2	40	11 JUL06	19 AUG06	INT02*
Integration and Test					
INT02	Integrate Units	30	20 AUG06	18 SEP06	INT03*
INT03	Test System	25	19 SEP06	13 OCT06	

Figure 4.10 Activities with successors may be dangling. Example with finish-to-finish successors

Activity ID	Activity Description	Orig Dur	Early Start	Early Finish	Successors
SRA Book Dangler Test					
Plan the Project					
PLN01	Plan the Project	40	01 JUN06	10 JUL06	UNIT101*
Unit 1					
UNIT101	Build Unit 1	40	11 JUL06	19 AUG06	UNIT201*
Unit 2					
UNIT201	Build Unit 2	70	11 JUN06	19 AUG06	INT02*
Integration and Test					
INT02	Integrate Units	30	20 AUG06	18 SEP06	INT03*
INT03	Test System	25	19 SEP06	13 OCT06	

Figure 4.11 Activities with finish-finish predecessors may be dangling

July 11, it begins before the Plan the Project is complete. This is a risky approach to building Unit 2 and probably not what the project manager wanted to happen.

One solution to this dilemma is to impose *both start-to-start and finish-to-finish* logic between Build Unit 1 and Build Unit 2 as shown in Figure 4.12.

The two links between Build Unit 1 and its successor Build Unit 2 will cause the correct result for the activity Integrate Units whether either the predecessor (Build Unit 1) or successor (Build Unit 2) is longer than its planned duration. Here are the two cases. In the first case, shown in Figure 4.13, the predecessor is 80 days long instead of 40 days. The Integrate Units activity is correctly pushed out to start on September 29 from August 20.

In the next case, shown in Figure 4.14, the successor Build Unit 2 is 70 days long instead of 40. Instead of starting early (it cannot because of its start-to-start predecessor

Activity ID	Activity Description	Orig Dur	Early Start	Early Finish	Successors
SRA Book Dangler Test					
Plan the Project					
PLN01	Plan the Project	40	01 JUN06	10 JUL06	UNIT101*
Unit 1					
UNIT101	Build Unit 1	40	11 JUL06	19 AUG06	UNIT 201, UNIT201*
Unit 2					
UNIT201	Build Unit 2	40	11 JUL06	19 AUG06	INT02*
Integration and Test					
INT02	Integrate Units	30	20 AUG06	18 SEP06	INT03*
INT03	Test System	25	19 SEP06	13 OCT06	

Figure 4.12 Parallel activities with both start-to-start and finish-to-finish logic

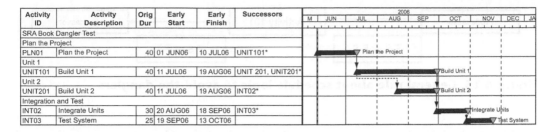

Activity ID	Activity Description	Orig Dur	Early Start	Early Finish	Successors
SRA Book Dangler Test					
Plan the Project					
PLN01	Plan the Project	40	01 JUN06	10 JUL06	UNIT101*
Unit 1					
UNIT101	Build Unit 1	40	11 JUL06	19 AUG06	UNIT 201, UNIT201*
Unit 2					
UNIT201	Build Unit 2	40	11 JUL06	19 AUG06	INT02*
Integration and Test					
INT02	Integrate Units	30	20 AUG06	18 SEP06	INT03*
INT03	Test System	25	19 SEP06	13 OCT06	

Figure 4.13 No danglers—right answer when predecessor is longer

Activity ID	Activity Description	Orig Dur	Early Start	Early Finish	Successors
SRA Book Dangler Test					
Plan the Project					
PLN01	Plan the Project	40	01 JUN06	10 JUL06	UNIT101*
Unit 1					
UNIT101	Build Unit 1	40	11 JUL06	19 AUG06	UNIT 201, UNIT201*
Unit 2					
UNIT201	Build Unit 2	70	11 JUL06	18 SEP06	INT02*
Integration and Test					
INT02	Integrate Units	30	19 SEP06	18 OCT06	INT03*
INT03	Test System	25	19 OCT06	13 NOV06	

Figure 4.14 No danglers—right answer when successor is longer

Build Unit 1), it correctly pushes Integrate Units to start on September 19 instead of August 20.[10]

We often run into parallel phases that are linked start-to-start and finish-to-finish, often with lags present, particularly in summary schedules where the phases are long and the preferred finish-to-start logic is just not available.

The conclusion from this exercise is that:

- Start-to-start logic alone devalues or may even cancel out the impact of elongation of, or risk in, the predecessor. The predecessor is a dangler.
- Finish-to-finish logic alone devalues or may even cancel out the impact of elongation of, or risk in, the successor. The successor is a dangler.
- This conclusion leads us to the general rule of predecessors and successors in dynamic CPM scheduling.
- Every activity except for the start milestone must have a finish-to-start or start-to-start predecessor to establish its beginning point.
- Every activity except for the finish milestone must have a finish-to-start or finish-to-finish successor that will be pushed by a later finish of that activity.[11]

A picture of this rule, shown in Figure 4.15, as applied to Activity 101 provides a useful reminder for the scheduler whose logic skills may need refreshing.

---

10   Of course better logic in this case would be to start both Build activities with the finish of the activity PLN01, Plan the Project and to start Integrate Units with the finish of the two Build activities.

11   The successor should be an activity that is pushed if that activity is late finishing. Tying the activities to the finish milestone is bogus logic and represents 'lazy scheduling.'

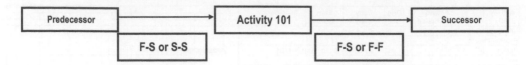

Figure 4.15 General rule about schedule logic to avoid the danglers

Many schedulers have never been taught this rule or they do not appreciate the need for the schedule to calculate correct dates and critical paths when durations change. This rule is important for CPM scheduling but it is crucial for the success of schedule risk analysis using Monte Carlo simulation.

Examine the Critical Path

The Critical Path is the central concept in CPM scheduling. The critical path is the longest path of activities linked to each other with logic through the schedule. The critical path determines the shortest duration of the schedule. The critical path is the path where the early dates and late dates are equal and hence total float is zero, where a 1-day elongation will cause the project to be delayed by a day. There are several definitions of the critical path.[12]

The critical path has many uses. Many feel that it is the path that should be most closely monitored and controlled since other paths have total float (total slack). A slack path may be late to the extent of its float but still not delay the project.

Many project managers and even some project schedulers do not look at their critical path. Some managers just ask what the completion date is without looking at how it is determined. Some schedules show the critical path running through the wrong activities, indicating some error in logic. Some critical paths even run through 'level of effort' activities, clearly a violation of project scheduling rules.

Some activities will have larger float than expected based on an understanding of the project. Some activities even have float in the high-hundreds or thousands of days on projects that are known to be under significant time pressure, but the project manager has not noticed this or does not see a great problem in those floats. One project manager once said: 'Control software is my problem.' The schedule had 200 days of float on software activities. He was not concerned when this was pointed out, and said: 'Well, that was before we knew how long software was going to take or where the activities needed to be hooked up. But, software is my problem.' He was not very concerned that the schedule did not show what he knew to be true.

If there are dangling activities or activities with incorrect logic their float values will not be accurate, the critical path may not be correctly identified and the completion dates may be inaccurate. Examination of the critical path is important to understand the schedule.

---

12   The PMI Practice Standard for Project Scheduling defines Critical Path as: The longest sequence of activities from the start date or the current data date of the project to the finish milestone, or the longest sequence of activities applying user defined point(s). (PMI 2007)

Incorrect Use of Lags

Lags are the device schedulers use to indicate that a successor activity may be delayed in starting or finishing until a certain amount of time has passed since its predecessor has finished or started. A lag is strictly for a fixed period of time that takes no resources and will not change. The example used is often 'watching concrete cure' but some use 'watching paint dry' or even 'watching coffee brew.' In detailed schedules where there are many short activities and finish-to-start is the predominant logical relationship, use of lags ought to be limited to durations between the finish of one activity and the start of its successor only when there is a specific reason, like the concrete is not cured, that the successor must wait.

In a summary schedule, some lags are permissible and may even be required by circumstance, although they should be used for the right reasons. For instance, the summary activities usually represent many more detailed activities that will not be specified until the detailed schedule is developed. In this case, it is not unreasonable to use lags instead of the detail that can not yet be shown. In most cases activities can substitute for these lags if there is any uncertainty in their durations.

An example of using lags on a summary schedule might be a construction project. There is a design phase and a procurement phase. As shown in Figure 4.16, some design time is required before any equipment can be put out for bid. Procurement can start before design is completed because the specification of the long-lead-time equipment is planned to start near the beginning of the design phase. A start-to-start relationship that has a lag to allow the specifications of long-lead-time equipment to be developed might be established between the start of design and the start of procurement activities. In this case there is also a finish-to-finish lag, indicating that fabrication and delivery of the last item designed takes some time.[13]

While there are some legitimate uses for using lags, many schedulers use lags in ways that are not appropriate. They use long lags to put successor activities on specific dates. In this case, dates have been established in some fashion ('back of the envelope') and the dates have been communicated to the scheduler. The predecessors and logic that forces the dates is not always known or the scheduler does not seek out the predecessor logic that would establish the dates by logic in the schedule. Sometimes predecessor activities are not immediately available to the scheduler.

ID	Task Name	Duration	Start	Finish	Predecessors
1	Start	0 d	6/1	6/1	
2	Design	50 d	6/1	8/9	1SS
3	End of Design	0 d	8/9	8/9	2
4	Procurement	70 d	6/29	10/4	2SS+20 d,3FF+40 d
5	Finish	0 d	10/4	10/4	4

Figure 4.16 Lags may be needed on summary schedules

---

13   This Microsoft Project® schedule shows how both start-to-start and finsh-to-finish relationships can be established between predecessor and successor using a milestone, End of Design.

What is a scheduler to do if management or a team leader says that Activity C needs to start on September 7? Failing to learn about the proper predecessor tasks and logic, the scheduler does what is easy and puts a long lag on an earlier task. Some schedules exhibit chains of start-to-start logic with lags, combining inappropriate use of lags with dangling logic. In many cases of lag abuse, the so-called schedule is really just a calendar on which some dates are specified. No constraints need to be employed, because the lags can be made to be of a size to place the successors on dates certain, at least in the baseline schedule. A picture of a typical such misuse of lags is shown in Figure 4.17.

ID	Task Name	Duration	Start	Finish	Predecessors	2nd Quarter Apr May Jun	3rd Quarter Jul Aug Sep	4th Quarter Oct Nov Dec	1st Qu Jan
21	Start	0 d	6/1	6/1		♦ 6/1			
22	Activity A	50 d	6/1	8/9	21SS	6/1 ▨▨▨ 8/9			
23	Activity B	70 d	8/3	11/8	21SS+45 d		8/3 ▨▨▨ 11/8		
24	Activity C	75 d	9/7	12/20	21SS+70 d		9/7 ▨▨▨ 12/20		12/20
25	Finish	0 d	12/20	12/20	24				♦ 12/20

Figure 4.17 Lags are inappropriate devices to set successors on dates

Using the lags, Activity B is placed on August 3 and Activity C on September 7. The bars look to be in alignment, but there is no logic to support it. The question is: 'What keeps Activity B and Activity C from starting now?' And: 'What happens to these activities if the Start is delayed?' There is no reason internal to the project, and the envelope, the back of which started this all, has probably been long thrown away. Used this way, a lag has no content; it is just a calendar, an admission of incomplete logic.

A lag cannot be analyzed for risk during a Monte Carlo simulation. It is fixed in duration (or in percentage with its predecessor activity). Since the lag has unknown content it would be difficult to put a three-point estimate on the lag in any case, even if the simulation package would accommodate different durations of the lag, which most do not do.

There is another issue causing lags to overstate the schedule risk during a Monte Carlo simulation. The (inappropriate) use of the lag to put a successor on the 'right' date gets the scheduler into trouble. The 'right' date only results before any changes such as statusing have been made to the predecessor activities. For instance, suppose the builder has informed us that the crew is not available until August 24. The scheduler knows that Design should finish on July 26, and figures that a 20-day lag will show the Build activity starting on the 'right' date, August 24. He produces the schedule in Figure 4.18.

ID	Task Name	Duration	Start	Finish	Predecessors	2nd Quarter Apr May Jun	3rd Quarter Jul Aug Sep	4th Quarter Oct Nov Dec	1st Qu Jan
8	**Before Status**	140 d	6/1	12/13		▬▬▬▬▬▬▬▬▬▬▬			
9	Start	0 d	6/1	6/1		♦ 6/1			
10	Design	40 d	6/1	7/26	9	6/1 ▨▨ 7/26			
11	Build	50 d	8/24	11/1	10FS+20 d		8/24 ▨▨▨ 11/1		
12	Test	30 d	11/2	12/13	11			11/2 ▨▨ 12/13	
13	Finish	0 d	12/13	12/13	12			♦ 12/13	

Figure 4.18 Use of a lag to set build on August 24

The minute a predecessor is extended, say by 10 days, the fixed lag pushes the successor off the 'right' date on to another date 10 days later. If Design were to take 50 days instead of 40 after statusing, the lag would cause the Build activity to start on September 7 as shown in After Status Using Lags portion of Figure 4.19 even though the builder could still start on August 24. From this example we see that use of fixed lags, which appear to be correct in the baseline schedule, may exaggerate the risk in the schedule by persisting when predecessors are delayed.

An alternative approach to this situation would be to use a start-no-earlier-than August 24 constraint on the Building activity. The builder has indicated that something outside of this schedule will lead the crew to be available for this project on August 24. (Is this true? Could there be a risk to this date from their other job?) Since the condition is outside of the schedule it is correct to insert a 'start-no-earlier-than' (SNET) constraint on the Build activity. Now if the Design activity is 10 days longer (or up to 20 days late) the Build activity still starts when the builder says the crew will be available, August 24, as shown in the lower panel of Figure 4.19 labeled 'After Status Using SNET'.

In summary, in Figure 4.19, when design is 50 days the lag approach does not work (see the 'After Status' schedule where Build is pushed out unnecessarily to September 7) but the start-no-earlier-than constraint works (see the 'After Status using SNET' schedule where build starts August 24).

The best alternative to using an artificial long lag between activities is to look for the predecessors and link them to successors with logic. In some cases these predecessors do not exist in the schedule and they will have to be discovered or created, such as the Design Review/Approval activity shown in Figure 4.20 that takes the 20 days previously supplied by the fixed lag.

ID	Task Name	Duration	Start	Finish	Predecessors	2nd Quarter		3rd Quarter		4th Quarter		1st Quarter		2nd Q
						Apr	May	Jun	Jul	Aug	Sep	Oct	Nov	Dec	Jan	Feb	Mar	Apr
8	**After Status using Lags**	**150 d**	**6/1**	**12/27**										
9	Start	0 d	6/1	6/1										
10	Design	50 d	6/1	8/9	9									
11	Build	50 d	9/7	11/15	10FS+20 d									
12	Test	30 d	11/16	12/27	11									
13	Finish	0 d	12/27	12/27	12									
14	**After Status Using SNET**	**140 d**	**6/1**	**12/13**										
15	Start	0 d	6/1	6/1										
16	Design	50 d	6/1	8/9	15									
17	Build	50 d	8/24	11/1	16									
18	Test	30 d	11/2	12/13	17									
19	Finish	0 d	12/13	12/13	18									

Figure 4.19 Use of a SNET constraint to set build on August 24 works better than a 20-day lag when the predecessor's completion date changes

ID	Task Name	Duration	Start	Finish	Predecessors	2nd Quarter		3rd Quarter		4th Quarter		1st Q
						Apr	May	Jun	Jul	Aug	Sep	Oct	Nov	Dec	Jan
1	**Before Status**	**140 d**	**6/1**	**12/13**								
2	Start	0 d	6/1	6/1								
3	Design	40 d	6/1	7/26	2							
4	Design Review / Approval	20 d	7/27	8/23	3							
5	Build	50 d	8/24	11/1	4							
6	Test	30 d	11/2	12/13	5							
7	Finish	0 d	12/13	12/13	6							

Figure 4.20 Find a predecessor (Design Review/Approval)

Finding or creating predecessors is preferable to using long lags to put activities on specific dates. The dates that are consistent with the static view will not be the 'correct' dates when predecessor durations change, and predecessor activities can be risked.

Lags often overstate the risk in a Monte Carlo simulation. In a simulation the durations change many, usually thousands of, times and the date will be 'incorrect' in most of these iterations. Once we introduce the Design Review/Approval activity, it can be examined for its risk and the simulation will be correct.

Ignoring Scarce Resources

Project activities are generally accomplished with the application of resources to tasks. Often resources are not sufficient to allow two or more activities to be performed simultaneously. The resources that are limited are called Scarce Resources and should be treated with some care in the schedule. The preferred approach is to:

- identify the scarce resources and include those resources in the schedule along with their limits;
- allocate the number of units of the scarce resources needed to the activities that need them;
- perform 'resource leveling' that keeps the resource from being over-allocated to simultaneous activities in any working period by delaying activities until the resources become available.

Most schedules found in practice do not include resources at all, or the application of resources is not consistent. For instance, some but not all resources may be identified and applied. Or, the application of resources is good in one area of the schedule but not in another. It is a rare schedule indeed that includes the full complement of resources correctly applied to all activities.

It is not difficult to identify the scarce resources and to allocate them to the activities they support, however. A project schedule that does not recognize the limitation of scarce resources and the impact that they may have on the ability to complete the project on time is not a realistic or reliable schedule. Figure 4.21 shows how a resource-limited schedule would look if resources were ignored. This project, that does not recognize the limitation that we only have one Test Station, finishes on 15 November 2006.

Activity ID	Activity Description	Orig Dur	Early Start	Early Finish	Resource	2006
Resource Leveling Test						
Start Milestone						
AAA	Start	0	01 JUN06	01 JUN06		
Unit 1						
B100	Build Unit 1	50	01 JUN06	09 AUG06		
T100	Test Unit 1	30	10 AUG06	20 SEP06	TEST STN	
UNIT2						
B200	Build Unit 2	50	01 JUN06	09 AUG06		
T200	Test Unit 2	30	10 AUG 06	20 SEP06	TEST STN	
Integration and Test						
I100	Integration and Test	40	21 SEP06	15 NOV06	TEST STN	
Finish Milestone						
ZZZ	Finish	0		15 NOV06		

Figure 4.21 Apply TEST STN to testing activities before leveling

Recognition that we really only have one test station leads to resource leveling. Resource leveling delays those activities that require the resources that are being used elsewhere. If we only have one Test Station, this project would be delayed until December 27 because the test station starts with Unit 1, then tests Unit 2 and finally the integrated system as shown in Figure 4.22.

Activity ID	Activity Description	Orig Dur	Early Start	Early Finish	Resource
Resource Leveling Test					
Start Milestone					
AAA	Start	0	01 JUN06		
Unit 1					
B100	Build Unit 1	50	01 JUN06	09 AUG06	
T100	Test Unit 1	30	10 AUG06	20 SEP06	TEST STN
UNIT2					
B200	Build Unit 2	50	01 JUN06	09 AUG06	
T200	Test Unit 2	30	21 SEP06	01 NOV06	TEST STN
Integration and Test					
I100	Integration and Test	40	02 NOV06	27 DEC06	TEST STN
Finish Milestone					
ZZZ	Finish	0		27 DEC06	

Figure 4.22 After resource leveling TEST STN

The realization that there is such an impact of resource limits on CPM scheduling indicates that we need to resource level each iteration of the Monte Carlo simulation as well. Each iteration must take account of the possible effect of resource scarcity so we can be assured that the dates and critical paths are at least feasible within the resource limits.

Ineffective Use of Milestones

Schedulers often represent important events in the schedule as milestones. Milestones are effective in representing the completion of a phase of the work or delivery of articles produced by the schedule's activities, including at the end of the project. Milestones are much less useful in representing inputs such as delivery of material, subsystems or software updates that are inputs to the project from third parties.

The supplier or subcontractor may promise to deliver the material or component to the project for further processing on a particular date. Schedulers often represent that date with a milestone, such as Component X Delivered or Software Version Y Available. That milestone is pinned on a date using a constraint such as 'must start on.'

The scheduler knows nothing about the inner workings of the supplier or subcontractor's processes and is not privy to their schedules. What they know is the promise date of the final result, delivery of the intermediate product. The supplier is often reluctant to status their progress for us. It is tempting to represent that date in our schedule with a start milestone.

What is a problem with this approach? These components or deliveries are promised by others and we have little understanding of the risk of the suppliers or subcontractors not making it on time. Yet, most project managers and schedulers have experienced late deliveries and consequential delays in a project. In a very real sense, promised dates, even if they are documented in a contract are risky. Using a constrained milestone to represent that date indicates much more certainty than is realistic. Some suppliers are known to be late habitually. In some circumstances, such as when they are very busy with many products to produce for different clients, almost all suppliers will be late some percentage of the time.

A risk analysis should be able to explore the possibility that deliveries from third parties may be late by including them in a risk analysis. Yet, we cannot put risk ranges on milestones in available schedule risk analysis software.

The better way to treat promised delivery dates that might be late is to use an activity, say, Supplier X Fabricates and Delivers Component A, that represents, at a summary level, the work of the supplier. The activity can have a start date that is the letting of the contract or purchase order and enough duration to place the finish date on the date delivery has been promised.

The two approaches to supplier promise dates—use of a milestone or a summary activity—each of which indicates that the project will be completed on October 11 if everything goes according to plan, are illustrated below.

In the project, Using Milestone for Delivery, shown in Figure 4.23, the milestone Delivery of Component X sits on September 6, the promised date. The amount of information that can be put on that milestone is just the date, not how long it takes or what risk there may be in its duration, and hence the delivery.

In the project, Using Activity for Fabrication and Delivery, also shown in Figure 4.23, the supplier's work of fabrication and delivery is represented by an activity, Fabricate and Deliver Component X. That activity can be reviewed and a risk probability distribution can be placed on it. Since it is an activity the project controls staff is more likely to try to get information from the supplier to status it.

There are two benefits to this approach:

1. The scheduler can see that important work is being done by others and may be able to status it in some gross sense with phone calls or visits to the supplier to get a sense if the work, and hence the delivery date, is on track.
2. A risk analyst can put a probability distribution on the activity representing the organization's assessment of the degree of certainty in this delivery.

ID	Task Name	Duration	Start	Finish	May	June	July	August	September	October
1	**Using Milestone for Delivery**	**95 d**	**6/1**	**10/11**						
2	Start	0 d	6/1	6/1						
3	Design	30 d	6/1	7/12						
4	*Delivery of Component X*	0 d	9/6	9/6						
5	Site Work	20 d	7/13	8/9						
6	Construction	20 d	8/10	9/6						
7	Install Procured Item	10 d	9/7	9/20						
8	Commissioning and Start-Up	15 d	9/21	10/11						
9	Finish	0 d	10/11	10/11						
10										
11	**Using Activity for Fabrication and Delivery**	**95 d**	**6/1**	**10/11**						
12	Start	0 d	6/1	6/1						
13	Design	30 d	6/1	7/12						
14	*Fabricate and Deliver Component X*	40 d	7/13	9/6						
15	Site Work	20 d	7/13	8/9						
16	Construction	20 d	8/10	9/6						
17	Install Procured Item	10 d	9/7	9/20						
18	Commissioning and Start-Up	15 d	9/21	10/11						
19	Finish	0 d	10/11	10/11						

Figure 4.23 Use summary task for supplier's promise date

Representing the third-party work of fabrication and delivery as an activity allows the risk analyst to investigate the uncertainty that will, potentially, affect the overall project.

Summary—Needed, a Good Project CPM Schedule

Project schedule risk analysis starts with a good project schedule. That schedule must be available, first of all, and represent the current plan of the project.

The first job of the schedule risk analyst is to examine the adequacy of the project schedule. Experience shows that the schedule risk analyst cannot leave this task to the project schedulers and that repairing the project schedule may take days before the schedule is ready for simulation.

All too often the schedule that is presented for schedule risk analysis is inadequate for the analysis. The main reasons include that the schedule does not really represent the project plan and that there are weaknesses in the scheduling discipline that keep it from providing the correct dates and critical paths when undergoing a Monte Carlo simulation.

There are several reasons that we find schedules inadequate for the task of schedule risk analysis:

- Scheduling is a difficult discipline. Many schedulers have not learned the discipline, have forgotten it, or are not given a chance to practice proper scheduling. For this reason, schedulers must have good technical skills.
- Gathering schedule-related data is difficult and requires schedulers and team leads to work together. For this reason schedulers must have good people skills.
- Management and customers often want the project to be finished sooner than is reasonable and they put great pressure on the scheduler to shorten activities to improbable or impossible durations and to schedule sequential activities in parallel. The scheduler is usually not in a position to resist this pressure, so the schedule may be incorrect to begin with.
- The rules of the scheduling discipline are precise but are not always known to practitioners. For instance:

  - Constraints can be used artificially to keep the schedule in the computer from overrunning, often while the project itself is overrunning its schedule. In some situations, these late-date constraints can bias the risk results.
  - Logic requires that activities have both finish-to-start or start-to-start predecessor(s) and a finish-to-start or finish-to-finish successor(s). This rule is correct; the well-known rule that all activities need a successor is not sufficient.
  - Critical paths and total floats should be examined to discover weakness in logic.
  - Lags are to be used in only limited and well-understood circumstances. It is likely that there will be more lags in summary schedules than in detailed schedules. Often lags are inappropriately used instead of logic, either to place activities on specific dates or instead of defining the activities that would fill the lag time.

- Scarce resources can be important in determining the progress of the project. A schedule, and the risk analysis of it, needs to take into account the scarce resources if they exist by leveling the schedule and the simulation.
- Milestones should not be used to stand in for dates promised by suppliers and subcontractors. They assume success of those third parties when success is usually not assured, and supplier uncertainty cannot be applied to zero-duration milestones.

The schedule will never be perfect. Most schedules need to be improved for risk analysis. The schedule must be improved before the risk analysis begins, but improvement may continue during the risk interviews, and sometimes even after the simulation shows some schedule-related anomalies. Once we have a good schedule we can be generally assured that the simulations provide the correct information under Monte Carlo simulation.

References

Hulett, D. (2007). 'The Problem with Dangling Activities.' *Cost Engineering* 49(1): 19–23.

PMI (2007). *Practice Standard for Project Scheduling*. Newtown Square, PA, Project Management Institute.

5 *Collecting Risk Data: Exploring Methods and Problems*

Introduction

The quality of the schedule risk analysis involves two main components. The first is the quality of the project schedule (see Chapter 4). The other is the quality of the risk data used in the analysis. In one view the quality of the data is the more difficult to ensure and the challenges more daunting. Surely the greatest amount of time in any risk analysis exercise is spent gathering good quality data necessary for deriving good quality results.

In one way the data used in risk analysis is not different from that used in any other planning or analysis function such as scheduling. Both involve estimating or predicting uncertain events since they require people to make estimates about the future. In another way, however, risk analysis seems to differ in character from, say, project scheduling or cost estimating, since it asks people to admit and calibrate their uncertainty about those estimates.

It is a puzzling fact that people seem to be more comfortable making single-point estimates of activity durations rather than providing an estimate of an uncertain future within a range. It is arguable that it should be easier to estimate using a range rather than a deterministic value. Experience shows that many people have very little experience with, or feel uncomfortable when expressing the degree of, their uncertainty.

Some people avoid considering project risk analysis entirely because they believe the data collection is so problematic as to be just guessing. Other people use the difficulty of collecting risk data and the paucity of risk databases from which to draw as their rationale for avoiding the exercise entirely, when, in fact, they just do not want to quantify risk on their project. Talking about risk is one thing, while putting numbers to those problems and fears can be another thing entirely.

Because the risk data are difficult to collect, we have to be both diligent and persistent in our efforts to gather the data and a little modest about their accuracy. Collecting quantified or calibrated risk information is a challenge. Still, experience tells us that valuable and accurate results can be generated in project risk analysis. As early as 1973 the US Air Force Risk Analysis Study Team concluded: 'Initial cost and schedule estimates for major projects have invariably been over-optimistic. The risk that cost and schedule constraints will not be met cannot be determined if cost and schedule estimates are given in terms of single points rather than distributions.' They concluded: 'A formal risk analysis is putting on the table those problems and fears which heretofore were recognized but intentionally hidden.'(US AirForce 1973).

Types of Risk Data Needed

Risk analysis explores uncertainty in important project metrics. The main data that help in quantifying the schedule risk are listed below:

- In schedule risk analysis the main issue is uncertainty in the activity duration (see Chapter 2), called the 'impact ranges' for each activity. These are often estimated directly as duration ranges, usually calibrated as three-point estimates. These ranges summarize the impact of all of the risks that may affect the activity's duration.
- In a later chapter (Chapter 8) we will introduce the concept of Risk Drivers and the notion of individual risks' impact on activity durations. This concept requires that risks' impact on activity durations be specified as a multiplicative factor. Risk drivers or factors are one step more basic than three-point estimates on an activity's duration.
- Probabilistic branching (see Chapter 7) and the risk driver method (see Chapter 8) require calibration of the probability of a risk's occurring on our project.
- Coefficients of correlation between activity durations (see Chapter 10) are often calibrated.

These are the main types of risk data needed for mainstream schedule risk analysis.

The impact ranges are often misinterpreted by interviewees, particularly by those interviewees who are inexperienced in the risk analysis process. Risk impact range estimating often starts with the schedule duration. Estimates of activity durations have generally three types of uncertainties that must be included when considering how the project plan could be affected:

- *Uncertainty based on estimating error.* This type of uncertainty, sometimes called 'ambiguity,' is well known and discussed in estimating class. Estimating error arises if there is a lack of definition of the work or some uncertainty in the input data needed to make the estimation. Estimating error is larger in the early phases of the project when the maturity of the estimate is less than it is later when engineering has progressed and information specific to the project has been collected from contractors and suppliers. For instance, conceptual estimates are often subject to wide ranges, say -25 percent and +50 percent, whereas definitive estimates are narrower, for example, -10 percent and +20 percent, since they have the benefit of ongoing detailed engineering.[1] Ultimately, before the project execution the owners' estimators have bids from contractors and a contractual basis of estimate. Even at that point the actual costs may be thought of as possibly falling within a range of -5 percent and +10 percent from the estimate.
- *Uncertainty based on variability.* Some uncertainty in activity durations is the result of a lack of knowledge related to conditions that are always present. Such conditions would include 'labor productivity' which is always present but may be incorrectly estimated. The presence of labor productivity is not in question, although the value that is appropriate to assume for a specific workforce or task in the project is unknown.

---

1   Notice that these uncertainty ranges due to estimating error show more probability of overrunning than under-running the estimate. This is fairly well-accepted practice.

This risk would be associated with a probability of 100 percent and a range of values or impacts on whichever activities are affected by this risk.

- *Uncertainty based on discrete risk events.* This uncertainty is typically about risks that are discrete events, with a probability of occurring that is less than 100 percent (and greater than 0 percent) and an impact range that may drive the actual value (the activity duration) away from the estimate. The impact of these discrete risks, should they happen, is often not symmetrical. Risks in project management are dominated by threats to the project that will cause more overruns than under-runs if they occur. Of course the opportunities that are found will also be skewed but in the direction of shorter durations. In addition, a risk range that includes both opportunity and threat possibilities is usually skewed more toward bad results than toward the good results.

Some Considerations that Make Gathering Risk Data Difficult

We have to be clear in collecting data that all types of risks are considered. Some inexperienced interviewees will provide estimating error only, either under the (false) impression that estimating error only is what we are asking for or perhaps because they do not have the experience or the comfort to provide variability and discrete risk impact ranges as well.

Several factors contribute to making the gathering or generating of project data about risk more difficult than the traditional data such as schedule durations and line-item costs, construction hours, prices per ton of steel or other data used to build up the project plan. These factors can be classified into those that pertain to individuals and those that relate to the cultural environment within which the risk analyses are conducted.

Impact of the Organizational Culture on Risk Data Collection— Motivational Bias

Organizational culture tends to set up penalties for questioning, challenging or just going against the corporate position expressed in the project plan and schedule. These penalties often establish an atmosphere where 'motivational bias' occurs. Discussing risk openly and realistically often has negative implications for the organization and those who give the risk data may be punished or otherwise made to feel uncomfortable or that they are 'not a team player.'

- The organization may have made representations to the public, lenders or government agencies about a specific number. For instance, suppose a company applies for and receives a project development loan for $100 million dollars based on a cost estimate and schedule for the project. If the risk analysis might call into question the cost and schedule that was used as the basis for the loan, there will be resistance and reluctance even to consider quantifying the project's risk. If the risk analysis is conducted the organization will exert pressure to make sure 'there is no risk in our plan.' If the risk analysis shows there is substantial risk it is likely to be thrown away and those involved may be punished.

- Management or customers may have decided that a project should finish on a certain day and that its cost should be a certain amount. Often these schedules and cost estimates have not been vetted with the experts and are in many cases inconsistent with unfettered realistic expert opinion. Contemplating a risk analysis result is not comfortable for the team because it may challenge management's target milestones.
- Any manager, team member or other Subject Matter Expert (SME) who is asked about the risk in the project is being asked, essentially, to challenge the very basis of the project plan. The implications of any honest risk analysis are unlikely to make the project look better, and most people do not look forward to being part of that analysis.
- There may be an atmosphere that is hostile to project risk analysis because it is new or unconventional, maybe even considered too difficult for the organization to perform. Often organizations do not want to look into risk. Anyone within the organization who raises the spectre that risk may occur and jeopardize the plans may be ostracized or punished in other ways. The term 'shoot the messenger' is common for the way these people are treated by a hostile corporate culture.

These organizational culture issues can trump any individual's desire to conduct risk analysis. Maj. David Christensen reviewed the experience from a US Defense Department acquisition program and concluded: 'Expert judgment must not be impaired by a culture that suppresses truth … In short, a so-called 'shoot the messenger' culture will absolutely destroy responsible decision making by biasing the database or the analysis of it.' (Christensen 1993).

The organization may be worse than unaware of project risk; it may actually be hostile to the concept. A hostile culture may be way beyond a simple lack of maturity in risk management. Aversion to project risk management is often strongest in middle management who are trying to make the project look good to management and other stakeholders. Sometimes top management understands that there is risk and would like to know that the project manager is aware of it and doing something about it. In other organizations the top management may also be opposed to risk management.

All of the risk management training, risk analysis software, risk management handbooks and desire by some to examine risk can be thwarted by a hostile corporate culture. Somebody who gets a salary from the organization may be unwilling to go up against such a barrier and jeopardize their career. For this reason alone it may be necessary to engage an independent expert from outside the company to conduct the analysis. An outside independent expert can be hired for a specific job and can often do that job honestly without jeopardizing their future or reputation. If a champion hires this consultant and supports them in the face of the corporate culture the job can be done with honesty and integrity. Experts do not depend on one organization for a large part of their income and often are hired in part based on their integrity as well as their expertise.

The good news is that actively hostile corporate cultures are encountered less frequently as time goes on. There are several factors that make the environment more conducive to risk analysis:

- Experience with the discipline. Sometimes just having someone in the organization, the 'champion,' conduct risk analysis and derive benefits will serve as an inducement and cover for others to follow.
- Organizations are adopting procedures and publishing handbooks for the conduct of project management. This often includes risk analysis, although quantitative risk analysis is not usually the first discipline to be included.
- Customers are demanding risk analysis for reluctant contractors. The problem with this approach is that the contractors have most of the information and the customers do not have their own experts ready to step in with the data.
- Government agencies such as the Department of Defense, Office of Management and Budget and the Government Accountability office in the USA have required risk analysis.[2]

Data Collection Problems that come from the Individual— Cognitive Bias

Risk data are inherently different from standard project planning data since they explore the limits of our uncertainty. While it seems counterintuitive, it appears to be easier for people to be definitive about a single number, say the duration of an activity, than to express a range within which that duration will probably fall.

People are often more comfortable in making single-point or deterministic estimates even if they know the number is likely to be wrong, perhaps because they learned in school or were taught on the job by people who were not clear about uncertainty. Of course, it is often useful to start with the uncertainty of the estimate components (for example, based on estimated hours of work, resources available, productivity of those resources and confounding factors) to understand the uncertainty in the duration estimates themselves.

Exploring the degree of uncertainty makes people uneasy because:

- Describing a range of values is tantamount to admitting that a scheduler does not know the answer. Most engineers, as a profession very influential in developing project plans and estimates, have a preference against ambiguity and in favor of precision. This preference is powerful and persistent, even when their own history provides examples where the exact estimates have been proven wrong. Individually they like to be definite, and suggesting that a value is correct only within a range is against their training and mindset.
- Uncertainty concepts such as the most pessimistic, most optimistic and even the most likely duration values are new to them. For some reason many people find that making a set of assumptions for a single-point deterministic estimate is easier than making a set of assumptions for the pessimistic or optimistic values and for examining whether the estimate is actually their most likely value.

---

2   'The program's milestone schedule should also be adjusted for risk. Measurable WBS elements significant to a project milestone should be analyzed for most optimistic, most pessimistic and most likely durations. A risk adjusted schedule will have finish dates that reflect the likelihood of a risk event occurring and its associated schedule impact. If schedule delays will effect cost, this information should be reflected in a risk adjusted cost estimate.' USOMB (2006). Capital Programming Guide OMB.

- Some other risk concepts are just plain new to team members. For instance, often they have not been asked to estimate the probability that a risk will occur. The question comes at them without any context so they are uncomfortable making an estimate. Calibrating the correlation between two activity durations is even more unfamiliar and puzzling to most people. If the analyst uses the normal distribution that is specified by its mean and standard deviation, they will probably find that estimates of the standard deviation will be essentially impossible to collect from most interviewees.
- Team members know that projects could proceed differently from the plan but each individual has a limited set of experiences from which to draw. Team members typically find it difficult to generalize from a subset of experiences to the broader concepts.
- Team members do not have training or mentoring in specifying the optimistic and pessimistic extremes. Their training usually revolves around how to make single-point deterministic estimates without any uncertainty (representing unreasonable hope in the face of uncertain reality). Even accepted 'fudge factors' or contingency reserve estimates are prescribed by management or their profession as exact percentages of specific line items or summary below-the-line accounts.

Individuals typically develop risk data by using their expert judgment. It is well documented that most people rely on heuristics or rules of thumb to develop these data. If heuristics were perfect the data would be very good. However, there are biases associated with the use of heuristics that are documented in the literature and in practice. These are generally classified as 'cognitive biases' and are discussed below.

Individuals can overcome the limitations that come from their own background or mindset in most cases by positive experience, expert facilitation and encouragement from management. Practice will help them overcome some of the internal barriers to providing risk ranges. Positive feedback and encouragement when they venture into the new concepts will help them overcome the resistance they initially feel.

However, the organization is not always supportive of the risk analysis exercise.

Risk Data are Usually not Available from History so We Use Expert Judgment

People attempting to develop plans for a project tend to be more comfortable if they have historical data that can be applied to future activities. They are comforted if there is a project that is recent and relevant to the one being planned. People are comfortable using historical data, perhaps adjusted for size, technical difficulty, weight or complexity, to help estimate future work. Even building up a detailed engineering estimate from scratch is made more credible if there is a database from which to draw in making estimates.

Unfortunately, most risk data are not available from historical databases, industry studies or recent and relevant projects. Of course a lot of specific activities in any project the contractor will bid on are familiar to that contractor or else they would not bid. Still, even historical data may be wrong as applied to a specific project:

- New projects are often different from old ones in ways that make estimating durations tricky even for semi-familiar work. In development projects where much is new most project managers feel very uncertain about any schedule or cost estimate. Often contractors will propose on a time-and-materials basis rather than fixed price basis in recognition of the impact of uncertain new activities on any plan or schedule.
- The impact on estimates of 'new information' or 'more detail' may make the application of historical data less relevant than it could be if information or detail were known for the new project. New information usually leads to an increase in the work from that estimated at a higher or strategic level, meaning that activities will end up longer than initially scheduled.
- Since 'hope springs eternal,' many contractors will hope, believe or claim that their performance on a new project will be better than it has ever been because of 'learning.' Historically, learning curves have been based on multiple (thousands) of repetitions of the same work (building airframes during a war) and cannot credibly be claimed after one or two similar projects have been completed. Even Lessons Learned programs may be less than optimally useful if not implemented carefully.
- The project schedule may be developed according to the customer's or other stakeholders' desires even if those desires are unlikely to be met in practice. Usually the dates are earlier than a professional estimate would produce.

Judgment Using Heuristics and the Biases that can Occur

When people are asked about the probability that something will occur or the range of impacts that might occur they often turn to heuristics or rules of thumb in building their answers. Use of heuristics is often beneficial since individual judgment may not have the benefit of relevant databases, studies or reports.

Research and experience has indicated that use of heuristics as a basis of reporting data about uncertain events often contains some well-known biases that should concern risk interviewers. Describing and understanding the better-known biases we may encounter during risk discussions and interviews tell us what to look for and, potentially, what type of questions we need to ask or adjustments we need to make to the data. Project risk analysis data collection is usually based on expert judgment and is subject to biases of representativeness, availability and adjustment and anchoring. A classic article in this discipline was written in the early 1970s and forms the basis of this section. (Tversky and Kahneman 1974). The discipline has grown and developed (Gilovich et al. 2002).

Representativeness Bias

Representativeness includes evaluating the probability or impact of an event by referring to another similar event. This is not in itself a bad thing to do—in fact we hope to have some comparable activities to use in judging the project we are now analyzing. There are some representativeness biases to consider, however:

Insensitivity to the prior probability of outcomes can apply when project leaders claim that their project is representative of successful projects even though there is a preponderance of project overruns in the company, industry or in their own past. These leaders discount or ignore the probability that this project may overrun like most of the earlier projects and believe (or at least claim to believe) that nothing bad will happen on their project even though it has happened on others. Examples might include:

- Many project managers indicate that 'this time there will be no design changes' even though in every other project design changes were common and time-consuming. Given the large number of projects where design changes occur it is almost right to say that the commitment to avoiding design changes may actually be irrelevant.
- Another example is the tendency of early estimates of cost and schedule to grow as more information becomes available and more detail is worked out. Even though the large majority of projects experience cost and schedule growth, a project manager may claim that it will not happen on this project or be angry when it occurs. The prior probability of cost growth is ignored when the current project is your own.
- Incorrectly weighing the influence on the project of prior probabilities and current events is fairly common and may be considered as an example of 'denial.'

Insensitivity to sample size is a very common bias observed during risk interviews. When asked about the probability of a risk's affect on a specific activity or about the range of a specific activity's risk from pessimistic to optimistic, many team members will apply the range based on the average over many different projects or from a large sample of activities. Using these ranges for an individual activity can be much too low for the probability or too narrow for the impact range on any specific project.

A practical example of this bias is seen when some interviewees make what could be called the 'error of composition' by ascribing an overall project risk range to each individual activity. These individuals are usually unclear on how risks of individual activities are combined to make overall project risk. If they are told that the overall project (or phase) has a risk range of '-20 percent to + 40 percent' they think that each component activity has that range as well. This is not true, of course. If each individual activity has a '-20 percent to +40 percent' range the range of the entire project will be substantially smaller in percentage terms unless the activity durations are perfectly correlated. Even though the actual range will grow, the percentage range declines as more and more activities are included in the complete schedule calculation because activity durations that are not perfectly correlated will not all be long or short together on a project—there will be cancelling-out as the project work proceeds down the schedule's paths. To achieve an overall risk range of -20 percent to +40 percent it would not be surprising if individual activities would need ranges of twice or three times those amounts.

It seems to be difficult for most interviewees to realize the width of the risk impact ranges that can occur on individual activities or risks. But, starting with narrow ranges for individual activities will result in underestimating the overall project risk. An example of this phenomenon is found in a later section of this chapter entitled 'A Suggestion to Address Narrow Ranges Provided During the Interview.'

The illusion of validity occurs when people predict the outcome that is most representative of the inputs with little or no regard for the factors that limit predictive accuracy. Unwarranted confidence may be placed in the validity of the project plan in the face of evidence of serious project risks. This confidence is the continual reliance on planning as a success indicator despite repeated demonstrations of its inadequacy. This illusion persists even among experienced project managers, attesting to its strength. The purpose of this observation is not to disparage planning *per se* but rather to be realistic about the benefits of planning and the perils of execution even after a plan is in place.

Availability Bias

The availability bias occurs when people assess the frequency of an event by the ease with which an historical precedent comes to mind. Risk data interviewees come into the interview with their own experiences and, when asked to quantify a probability or an impact range, they delve into their own memory for relevant comparisons. It is good to refer to your own experience or that of others, but in doing so some of the events brought out for comparison have more importance or impact on the answers than they warrant. These are the historical events that are easy to remember, perhaps because they were dramatic, made an impression at the time or had major consequences.

The availability risk is when a past experience used as a benchmark is affected by factors other than its frequency or probability. Notably the familiarity of a particular event or the dramatic impact the reference event has on the interviewee may make a relatively unlikely or dissimilar event come to mind. If this happens, the assessment of risk on the current project may be biased in the direction suggested by the dramatic or familiar event as shown in Figure 5.1.

An example of the availability bias might be a failure of a project component on a prior project. Suppose the interviewee was involved in a prior project where their assigned component failed and they were held up to criticism. Any project that uses that component might go to this person as a SME. The interviewee's initial assessment is that the component is very likely to fail, since it is easy to remember when that happened to them. The interviewee is very likely to insist that the probability is high for failure on the current project in the face of argument from others. This is one bias that may lead to an overestimate of project risk.

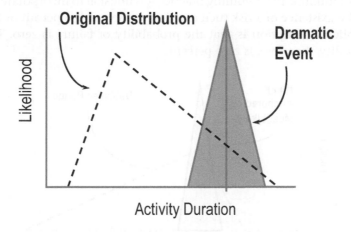

Figure 5.1   Availability bias can lead to an overestimate of risk

Adjusting and Anchoring Bias

Anchoring is expressed when people start from initial values that are taken as a basis from which a value to be estimated can be adjusted. In practical terms this anchor is often represented by 'the activity duration' found in the baseline schedule. The initial baseline value takes on a power, almost like gravity. When someone is asked to estimate the optimistic and pessimistic values the adjustment from the baseline estimate is often insufficient. When the interviewee is asked to report on the 'most likely' duration they will often refer to the baseline plan value even if, in another breath, they state their belief that the schedule is 'aggressive' or 'optimistic.'

Research shows that the anchor need not even be a valid point of reference to have the ability to bias the adjustment. As with other heuristics, adjusting from a starting point is probably a good approach when estimating values using expert judgment. However, there is evidence that people tend to make insufficient adjustments when doing so. In the risk arena this leads to underestimation of project risk.

If the interviewee has had a hand in making the schedule estimate, that estimate has additional strength in pulling the adjusted extreme optimistic or pessimistic ranges back toward the estimate. Figure 5.2 illustrates the underestimation of duration risk that often occurs in these interviews.

Another example of adjusting and anchoring is the attempt by interviewees to estimate the probability of a threat event, which is a risk event with negative consequences. Estimation of this probability can take at least a couple of forms:

- Probabilistic branching (see Chapter 7), which is often used to account for the possibility of failing a test leading to several new activities, requires collecting data about the likelihood of failing a test.
- The risk driver method of determining the schedule risk from the risk events usually found in the Risk Register (see Chapter 8) is another instance where we collect data about the probability of a risks' occurring.
- In normal planning and scheduling practice it is unusual to incorporate in the baseline schedule the existence of a risk such as a delay in securing a permit or failing the test, so the implicit assumption is that the probability of failure is zero. The anchor for that probability, therefore, is zero percent.

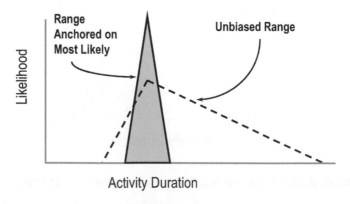

Figure 5.2　Picture of adjusting and anchoring bias producing narrow risk ranges

In order to estimate any positive probability for such an event the interviewee often starts at zero, the value implicit by the absence of failure activities in the plan or schedule, and adjusts to a positive value. While the adjusting and anchoring bias is in play there is another factor tugging the probability value toward zero—the fact that failing a test or not getting a permit is perceived by the interviewee and the project stakeholders as a real problem. Admitting that the problem has a probability of occurring, and that the probability may be fairly large, may have negative consequences for the project and the interviewee's future with the company.

What is the interviewee to do when there are two active factors forcing the estimated probability toward zero? While the probability of failing a test may indeed be small, the interviewee often underestimates the probability of the event.

Tversky and Kahneman report on some research indicating that people overestimate the probability of 'conjunctive events' as part of the adjustment and anchoring bias. A conjunctive event occurs if some event, such as finishing an activity within the scheduled duration, must happen over and over again to produce a successful project. Put another way, if any activity in the critical path takes longer than its scheduled duration the path and hence the project will be late.

In risk interviews we often hear: 'Everything has to go right for this schedule to be successful.' That statement indicates that the durations are aggressively estimated and the speaker actually expects one or more activities to exceed the plan. Even though the probability of any one activity's taking longer than scheduled may be low, with enough activities the probability that at least one activity takes too much time can be fairly large.

Tversky and Kahneman say:

> The successful completion of an undertaking, such as the development of a new product, typically has a conjunctive character: for the undertaking to succeed each of a series of events must occur. Even when each of these events is very likely the overall probability of success can be quite low if the number of events is large. The general tendency to overestimate the probability of conjunctive events leads to unwarranted optimism in the evaluation of the likelihood that a plan will succeed or that a project will be completed on time … Because of anchoring people tend to underestimate the probabilities of failure in complex systems (Tversky and Kahneman 1974).

Using New Information to Modify Prior Judgment

Often the project team will be given new information that conflicts with their prior assessment of the project. New risks may appear as the project progresses and they often conflict with the initial assessment made when the project was approved that it is on-plan. How should this new information change the prior assessment of the project's risk? There are three general positions that one can take:

1. The new information is the only thing that matters. Prior judgment has been shown to be wrong by more current events. This is thought of as a 'panic mode' response.

2. The new information is considered to be irrelevant since the funding decision has been made and even communicated to others in the baseline plan. This response is similar to the 'see no evil, hear no evil, speak no evil' mentality.
3. The new information is given its due weight, along with the prior assessment that was made with the same care and attention based on information available at the time. This response is generally the measured and mature response.

As a common example, suppose a project is viewed to be low risk and is approved on that basis. Then, some problems arise that were not anticipated. The project manager needs to evaluate those new events and see what they represent for the project at hand. The common interpretations of this event include:

- These specific problems observed represent a failed project with cost/schedule/quality problems. While this interpretation conflicts with the low-risk project as earlier assessed, it is a common reaction to problems that arise. Often people respond most to the last person to leave the office.
- These specific problems represent only the types of issues found in all successful projects and do not change the earlier assessment of a low-risk project. This represents ignoring more recent information because admitting its importance would require adjusting the prior assessment, changing the plan and perhaps even cancelling the project.
- Each of these interpretations by itself is probably wrong, and a proper blending of the two would be the right approach. Figure 5.3 shows how current pessimistic events might cause the prudent project manager to adjust their prior assessment that the project is going to be successful, leading them to change the position and shape of the probability distribution.

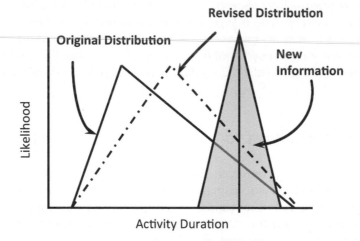

Figure 5.3   Adjustment of prior assessment about the project from new information

The proper weighting of the two factors, (1) the prior expectation and (2) the new information is derived from Bayes' theorems.[3] (Barclay 1977) Figure 5.3 indicates how new information, when added to the original distribution, might lead to a revised distribution of risk impacts on activity durations.

The Risk Interview: Collecting High-Quality Risk Data

Organizing and conducting the risk interview can contribute to the success of the risk data gathering exercise and enhance the credibility of the risk analysis results. There are several aspects to the risk interview. It is particularly important for those organizations and individuals that have low-risk management maturity levels and are inexperienced in risk analysis to be alert to these issues.

WHOM TO INTERVIEW?

Risk data are generally best collected in an interview session with people who are knowledgeable about the project or some aspect of it. Called subject matter experts, or SMEs, these people have experience in the industry of the project, its location, the markets in which its equipment/material/labor will be purchased, the regulatory environment surrounding the project, the technologies in use, the performing organizations and the customer or sponsor organization.

Most often the SMEs are members of the project team and its management, contractors or advisors. Collecting data from the team members has benefits and limitations:

- On the one hand, team members have the most familiarity with the project and have been intimately involved with the issues, assumptions, options and information about the project.
- On the other hand, the team members are often too close to the project. They have made the decisions and estimates themselves and may be suffering from some of the biases discussed above, particularly the anchoring and adjusting bias. In addition, since any negative news about the project could get it cancelled there is a motivational bias against making the project look questionable. Practical experience indicates that the higher in the project management hierarchy the interviewee is, the more likely it is to get data that are biased toward making the project look good.

---

3   Bayes' theorem relates the conditional and marginal probabilities of events A and B, where B has a non-vanishing probability:

$$P\ (A|B) = \frac{P\ (B|A)\ P\ (A)}{P(B)}$$

Each term in Bayes' theorem has a conventional name:

P(A) is the prior probability or marginal probability of A. It is 'prior' in the sense that it does not take into account any information about B.

P(A|B) is the conditional probability of A, given B. It is also called the posterior probability because it is derived from or depends upon the specified value of B.

P(B|A) is the conditional probability of B given A.

P(B) is the prior or marginal probability of B, and acts as a normalizing constant.

Intuitively, Bayes' theorem in this form describes the way in which one's beliefs about observing 'A' are updated by having observed 'B'. Wikipedia (2008). Bayes' Theorem.

Sometimes SMEs who are not on the project team itself but have relevant experience can provide less biased information. Often these people work in the organization where the project resides and therefore facts needed for the analysis can be revealed to them that may not be revealed to interviewees from outside the organization, for example, contractors. These non-team SMEs are generally supportive of the project and organization while they are more immune to the fallout of the risk analysis results than are project team members. For these reasons, in addition to bringing a new set of experiences to the interview, these non-team SMEs have less stake in the success of the project and can be relied upon to provide less biased data than those who are on the project.

ORGANIZATIONAL ISSUES CAN AFFECT THE QUALITY OF RISK DATA COLLECTED

If the risk analyst reports to the manager of the project that is being evaluated, then that project manager may bring pressure to bear, directly or indirectly, to make the risk analysis results favorable to the project. Often project managers want to avoid the risk analysis process entirely or, failing that, to make the results validate the approved project plan.

The project manager wants to go about the project business without having to adjust the plan or explain to management why the plan adopted is risky and what they will do to mitigate the risk. Often the project manager claims or actually believes that the project can recover successfully from any risk. The project manager wants to report good news to the organization and project stakeholders and hopes that upper management will let them alone. 'Just leave me alone and let me build my project.'

Risk analyses conducted by project managers often are a 'whitewash' of the project. These usually self-serving analyses are worse than not worth the effort—which may be very little—that went into them because they are misleading. Organizations that close their eyes to risk and incorrectly emphasize how good the project plan is will lose the opportunity to deal with risks. They are using 'project management by crossing your fingers,' usually a self-defeating strategy.

The organization that wants a realistic risk analysis result should adopt an organizational structure that ensures that the risk analysis is independent of the project manager. In some companies the risk analyst is organizationally located in the PMO or in a centralized corporate engineering function but does not report directly to the project manager. From that vantage point the analyst reports to a risk officer or high-level risk review board rather than to the manager of the project under analysis. Of course it is useful to present the analysis results to the project manager before submitting it in final form so that any errors can be corrected and the project manager is warned about the results. But, the analyst should be protected from repercussions at the hands of the project manager if they indicate the project will not finish on schedule by making them independent of that project.

Often an organization will hire an outside consultant who is expert in project risk analysis and who knows something about the industry to conduct the risk analysis. External consultants can be more objective than those who derive their entire livelihood from the project organization since that consultant:

• does not derive much of his income from any one company;

- will leave at the end of the engagement whatever the results;
- bases their reputation on their objectivity.

Still, even external risk experts sometimes face serious pressure from the project management hierarchy.

The level of the risk management maturity (see section below, Risk Management Maturity of the Organization) of the organization may be indicated in part by the degree to which the risk analysis function is independent of the project under review. In some very risk-mature organizations, risk analysis is standard practice and people are experienced in providing risk data so organizational separation may not be necessary. However, in many organizations creating a structure in which the independence of the risk analysis, and specifically risk data gathering, is institutionalized can remove a barrier to effectiveness.

CONDUCTING THE RISK INTERVIEW

Interviews to elicit project risk analysis data are of at least five kinds. Not all risk analysis exercise would have every type of interview represented:

- *Detailed information* is being sought about risks or schedule activities from people with specialized knowledge. These interviews are often focused on specific areas of the project such as procurement, design, construction, permitting or commissioning. These interviews may include one or a few individuals who are involved in or at least knowledgeable about those elements of the project.
- *Summary or overview information* is being sought, often from the project manager or others in the organization with a broad scope of responsibilities. These interviews may be most successful with the risk driver approach (see Chapter 8) where the basic building blocks are individual risks. Many of the risk register risks are broad, high-level risks that can be discussed from the '30,000 feet' perspective.
- *Background information* may be contributed by some interviewees. Some interviewees are not particularly familiar with the project. Information of a general nature can be helpful. Often, under expert questioning, someone who begins the interview saying they do not believe they can quantify any risks is soon able to quantify the data.
- *Confirmatory or contrary information* can be gathered by interviewing more than one person about each risk or activity. The benefit of having several experts quantify a risks' probability and impact is that some people will have a different perspective or opinion that should be considered. The drawback to having several opinions is that the analyst needs to be able to synthesize the data that will be used for the analysis from several sources.
- *Independent information* is sometimes collected from people not on the project or even not in the organization who are knowledgeable in the type of project, region of the world, technology, supplier and labor markets or regulatory situation. Participant experts from outside the organization will either sign a non-disclosure agreement (NDA) or may not be given proprietary information.

During the interview the interviewees are encouraged to bring with them any documentation that will help. People who bring in actual records from recent, relevant projects usually contribute better information because they can refer to real data. As an example, data on actual projects may be the antidote to belief in excessively-optimistic plans. Of course reliance on historical data has a downside since it may refer to only one or a few instances of the work under investigation and may not be representative of the true risk of the current project.

In many interviews the interviewee has no data to use as a reference and all of the heuristics and their biases discussed above can color the answers. The interviewer needs to be alert to the biases of the interviewees.

Risk interviews are more productive if the interviewees look at the risks and write down their preliminary probabilities and impact ranges ahead of time. If project teams prepare the data together the quality of the resulting data will benefit from those discussions. These interviewees or teams should bring to the interview filled-in tables of probabilities and impact ranges.

Experience shows that the pre-interview estimates of probability may be too low and of risk impact ranges may be too narrow, but it is very helpful for interviewees to have considered risk and prepared responses in advance. One of the interviewer's main responsibilities is to recognize when the biases affect the responses. Interviewers should probe and challenge the rationale of the interviewee to find out the full extent of the risks. Even carefully-prepared-in-advance probabilities and impact ranges often change during the interview.

STRATEGIES FOR INTERVIEWING ABOUT IMPACT RANGES AND RISK PROBABILITY

In asking about risk impact ranges, a useful technique is to address the pessimistic case first. Experience shows that the pessimistic scenario is the most difficult case for interviewees to quantify. We often ask the interviewee to first describe what events would lead to the worst-case scenario, then to estimate the value that it represents.

The pessimistic scenario may include several elements. The interviewer must be sure that it is at the edge of worst-case believability that the elements could all occur on one project. A useful device is to ask the interviewee to then make an estimate of the pessimistic case impact that reflects the scenario just described. Try to elicit the quantification of the pessimistic case where there is only 1 percent likelihood that the risk will be worse, since if you ask for a value that will never be exceeded the interviewee may spin some pretty bizarre stories that lead to values that are so extreme as to be useless in simulation.

The optimistic scenario should be explored next. Experience shows that people are generally not as aware of opportunities in the current plan (as distinct from risk mitigation that would change the plan) to make the project plan better as they are of threats to the project's objectives. The optimistic scenario is where opportunities should be listed. It is useful to get the description of the optimistic scenario before trying to quantify it.

Finally, for impact ranges we need to explore the most likely scenario. This is the scenario that has the highest probability, more probability than any other scenario, of occurring on this project.[4] People often think that this scenario is the duration for the

---

4   The most likely duration is the mode or modal value, the value that is most likely to occur. It is different from the mean, which is the weighted value of all possible values (each possible value is weighted by its relative probability of occurring). The mode is also not the median value, above which and below which 50 percent of the possible values lie. Only if the probability distribution is symmetrical will the mode be also the mean and the median value.

activity that is listed in the schedule. After all, wouldn't any reasonable plan be based on the most likely values? However, the most likely duration may not be the value in the schedule. In fact, many times the interviewee looks at the duration and designates that as the most optimistic value because the duration listed in the schedule may have been optimistic to begin with, or new data may have surfaced that make the old estimate obsolete. If the activity duration in the schedule is optimistic then the most likely duration is above that value in most cases.

The interview could also focus on the possibility of adding activities representing discontinuous events like failing a qualifying test, requiring new activities (such as identifying the root cause of the failure, planning the recovery, executing the recovery plan and retesting the article) that are not in the baseline plan. Risk events such as failing a qualifying test that lead to activities that are not usually included in the schedule lead to probabilistic branches (see Chapter 7) and can include significant schedule risk. If the probability of failing the qualifying test is high or the impact is large if the article fails the test, a significant amount of time of the interview should be focused on developing the activities of the branch, the probability of their occurring and their risk ranges.

STRUCTURE OF THE RISK INTERVIEW

Before the interview it is good practice to brief the participants, perhaps all in one meeting, about the data needed, the role and importance of the data to the analysis and of the analysis to the success of the project. The project manager or another stakeholder with some standing should introduce this briefing with a statement of how important the risk analysis exercise is to the organization, that we need to get the data for the success of the project and how much they personally support the exercise.

It is also helpful to provide the prospective interviewee with a questionnaire or form that will be filled in with information, such as probability and impact and listing all of the risks identified up to that point, during the interview. The interviewee is encouraged to attempt to complete the form before entering the interview and to bring the document with them.

The risk interview often begins with an introductory period during which the interviewee is getting settled for a productive discussion. This introductory period has several purposes:

- Overcome the interviewee's resistance to the process or reluctance to be interviewed. Sometimes the interviewee will be unfamiliar with risk interviews and may be suspicious about it, unsure of the process and generally resistant to providing risk data. The risk analyst needs to answer the interviewee's questions patiently and to reassure them of the process. The interviewee needs to participate willingly but that willingness does not come guaranteed with every interviewee.
- Answer questions about concepts. The questionnaire will ask for probability and/or impact ranges, and the interviewee may not understand what is needed. Answers that seem to be clear to the interviewee are not always what the questionnaire was getting at.
- Familiarize the interviewee with the basis of the risk analysis and its frame of reference. Answer any questions that pertain to the analysis.

- Inform the interviewee that their responses will be confidential and will not be attributed to them personally. Nobody will be told which specific person gave a specific response. The promise of anonymity is important to establish the atmosphere of confidence necessary between the interviewee and interviewer. It will help many interviewees to speak freely and honestly, knowing that there will be no repercussions for honest opinions. The risk interviewer should keep notes for the documentation, but those notes are not made available to the project manager or anyone else in authority over the interviewee. It is sometimes necessary to exclude the project manager or team leader from the interview in order to ensure complete candor from the interviewee.

- Ensure that enough time is available for the interview. Often people will assume that the interview should take only 30 minutes or an hour. Experience tells us that risk interviews may be that short with people who have only a few risks to discuss, have participated in risk interviews before, are comfortable with expressing their opinions about risk, and are economical with words. Often risk interviews can take 2 hours or more if the interviewee has a lot of risks to discuss, takes a while to get settled and comfortable, needs to familiarize themself with the process or just likes to talk. Lengthy interviews can be 2–3 hours and on very rare occasions have lasted half- to all-day. It is best to prepare the interviewee for a realistic interview time.

- Finish the risk interview in one session. It will be very difficult to get the interviewee or team back for a continuation session. We find that the inconvenience of the interview is less if it is allowed to go on until concluded. Do not be surprised if the interview runs over the initial allotted time. On occasion the interviewee becomes so involved or intrigued with the subject that the session lasts longer than it needs to, but in this case good quality data are usually received.

- Emphasize the need to look for opportunity-type risks. These are risks which, if they occur, would help the project stay on plan or improve on the plan. This emphasis on opportunities is not for the purpose of making the project manager happy. Rather, without emphasizing opportunities interviewees may not recognize them even if they are present. Most people are used to thinking that risk is a threat to the project's objectives. Interviewees and facilitators alike are generally not used to thinking of uncertainties that can help the plan. If we are not explicit about finding opportunities we will certainly miss them and the risk data will not be complete.

- Risk interviews often start out very slowly. The introductory period can take half an hour or even more. Even after the introduction, when risks are first discussed in substance, deciding on a value for the probability or impact range of a risk can take some time. It is not unusual that data on only two or three elements of risk data are collected by the end of an hour. At that point an interviewee may become discouraged since the progress extrapolates to a very long interview. However, participants should take heart! There are several reasons why the interview tends to proceed faster and faster as it goes along:

  - The introductory period is behind us and does not have to be repeated.
  - The risk concepts become clearer and clearer. Even interviewees who have never participated in a risk interview before will become more comfortable with the concepts as they go along.
  - When some risks have been successfully calibrated, succeeding risks may be easier

to deal with. For instance, later risks may be calibrated from risks discussed earlier in the interview since a later risk may be assessed as being more or less likely or having a wider or narrower impact range than a risk already discussed. Reference to risks that are discussed and calibrated earlier in the interview makes collecting data on later risks easier.

- It is not uncommon for people who are initially resistant to the entire risk analysis process to become enthusiastic and to buy into the process as the interview goes along. They may become more comfortable deciding on values for the risk probability and impact ranges as the interview proceeds.
- Some interviewees who enter the room believing that they can contribute nothing, and certainly cannot 'put numbers' to the risks may find out, with a little expert coaching by the facilitator, that risk calibration is easier than they thought.
- It is not uncommon for people who are new to risk interviewing to become adept within the space of their first interview. If interviewees have been involved in risk interviewing on other projects the interview will proceed even more quickly.

At the end of the interview the risk analyst-interviewer needs to document the data collected as quickly as possible. Putting the values and explanatory comments on a computer with the aid of a projector so everyone sees the recording of the interview can be that documentation. It is difficult to wade through days-old notes to try to remember what a specific interviewee said and why they said it. Recording the results, including the data and their rationale, in a simple spreadsheet database will help the interviewer decide on the values to use in the analysis when data gathering is done.

The Risk Workshop

Sometimes risk ranges are developed in workshops with 10–25 participants rather than interviews with one or just a few participants. Workshops can provide a good environment for sharing information and having a cross-disciplinary discussion. Risk data gathered in a workshop has the potential for being accurate because of the contributions of many people in the room.

The workshop usually starts with a preliminary discussion of the process that will occur including the type of data needed by the end of the day. From preliminary work, a list of relevant 'candidate risks' should have been developed.[5]

The workshop may be separated into smaller working groups, each led by a facilitator, with assigned risks to review. The risks can be a sampling of all risks in the list, so they cover many areas (engineering, construction, procurement, planning, financing, decision making, commissioning, permitting and the like) or they can be segregated by subject matter area. In either approach the participants are assigned to subgroups. We often need more than one expert facilitator to handle the different groups.

Gathering risk data using the workshop approach may suffer because of the possibility for intimidation. Workshop attendance often includes people at different grade levels

---

5   Often there is a risk register developed in advance by the team, but if it is not available the analyst is advised to develop one. This effort may require identifying project risks, perhaps using a risk breakdown structure, and prioritizing risks using methods based on assessing risks' probability of occurring and impact on the schedule if they were to occur. PMI (2004). *A Guide to the Project Management Body of Knowledge*. Newtown Square, PA, Project Management Institute.

in the organization. Sometimes the SMEs will be in the same workshop with their supervisors, the project manager and team leaders representing three or more hierarchical levels within the organization. The facilitator should insist that during the workshop there is no hierarchy, that everyone's voice is equal, and that all can contribute with valuable information, but some will not believe it. Even with these assurances, workshop participants at the lower levels of the organization may not be willing to say anything that may anger or contradict the position of their supervisor or the project manager for fear of being criticized or ostracized. The intimidation effect is greater in cultures where the custom of hierarchy is stronger. It takes a strong person at a lower level to introduce a threat-type risk into a workshop with their team leader, the project manger or other stakeholders present. This is why the guarantee of anonymity in the interview is so important and why interviewing with a single person or a few people without management being present is often more successful in gathering realistic risk data.

Frame of Reference for the Risk Analysis

A project risk analysis needs to have a frame of reference and that must be agreed to before the interview or workshop starts with the substance. The frame of reference sets the boundaries about which risks will be considered and which will not. The basis can depend on the perspective from which the analysis is conducted. The perspective is often different depending on whether the analysis is being conducted for the owner or a contractor, for instance:

- Whether change orders during software coding, equipment fabrication or construction constitute a risk depends on whether the analysis is being done from the owner's or contractor's side. The owner will probably consider change orders to be part of the project and the risk of change orders is included. The contractor will be paid for change orders so contractors only want to analyze risk associated with the contracted work.
- Risks over which the project manager has no control are still risks to the project. Often some interviewee will say: 'We cannot do anything about it, so it is not a risk.' This is incorrect. It could be argued that any risk that cannot be affected by the project team is more of a risk to the project than a risk that can be mitigated. Even if the project manager or owner cannot do anything about the risk they can position the project so that the risk, if it happens, has a smaller impact than if the project were not prepared.

Risk Management Maturity of the Organization

Initially the success of the project risk analysis depends on the risk-maturity of the organization within which it is conducted. Several aspects of risk management maturity are important (Salim and Hulett 2008; Hillson 1997; Hall et al.):

- People and resources must be available in sufficient quantity and trained to do the risk analysis. They should be expert in the field of risk analysis and have mastery of the tools including the people skills of the interview, the software skills to analyze the data and communication skills to present the results.

- Practices and standards must include project risk management as a 'best practice' that is expected to occur on all important projects. Risk management must be considered to be part of a successful project and its results must be incorporated into each decision node or stage gate review. Risk management that is considered an 'add-on' will be less successful than if it is routine practice.
- Leadership must be alert to the benefits of project risk management and supportive of it, rather than hostile or indifferent to it. Leadership should model risk management and be seen to use the results to choose between projects and make project success more likely.
- The mindset of the organization must favor discovering the reality of the risks rather than hiding risk. Honesty and objectivity must be rewarded, not punished by the organization.
- The organizational structure must be conducive to risk analysis. Risk data collection must be protected from the motivational biases that are common when project managers control the process and its results.
- There needs to be a network of risk management professionals who will both support each other and promote a dynamic and progressive view of best practices within the company. Sometimes outside expert consultants are important members of this network.

A Suggestion to Address Narrow Ranges Provided During the Interview

The bias of adjustment and anchoring leads to ranges that are too narrow. Also, because of the representative bias, people may take the impact ranges that apply to the entire project estimate and assign those ranges to the individual plan elements, again resulting in narrow ranges on individual activities.

For the impact range of a single activity or a risk we want to know the so-called 'outliers,' the very best and the very worst that may occur. Yet we are often given a narrow range. Team members may describe a serious risk but then quantify its impact as only -10 percent or +20 percent on the duration of the activity or risk being discussed. It appears to the interviewer that the range provided is too narrow given the risk description just provided. The interviewer becomes concerned that the interviewee is biased. Even under questioning by the interviewer, the interviewee may continue to insist on the ranges given. In another case, interviewees state as they are leaving the room that they provided P-10 and P-90 values. It is too late to reconvene them to collect the needed data on the extremes of the distribution, and they may be reluctant to give extreme values anyway. What should the experienced interviewer do in these cases?

The Problem When the Activity Ranges are Too Narrow

A practical example will serve to illustrate the problem with developing narrow ranges rather than wide ranges that can occur on a specific project schedule. Consider a 100-day project phase for which it is agreed that the worst case is 160 days and the best case is 80 days, for a range from 80 percent to 160 percent of the phase duration. (For this exercise we will assume that the most likely duration is the 100-day estimate.)

Now consider that this phase is made up of 10 activities of 10 day each. Interviewees often try to put the overall risk range (80 percent to 160 percent) on each of these shorter activities. Simulating the 10-activity path using 80 percent and 160 percent ranges for each individual activity provides the unreasonably narrow path result of from 95 days to 134 days. This result is shown in Figure 5.4.

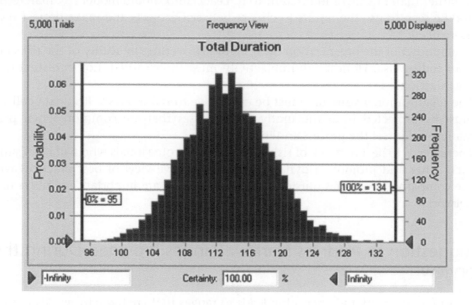

Figure 5.4   Path distribution is too narrow if the 10 activities have a range of from 8 days to 16 days

This result is much narrower than the range of 80 days to 160 days which was initially judged sufficient for the path as a whole. Why is that? It is because if the activities are independent (see Chapter 10) there is a lot of canceling out of randomly-selected durations on the short activities down the path. The Monte Carlo simulation software, representing the actual project, may randomly select a high value for one or even several activities but low or moderate durations for the others for a specific iteration. As durations are added down the path these values cancel each other out. There is no chance, in the simulation or in a real project, of getting a range from 80 percent to 160 percent if the individual activities only have those percentage ranges. The percentage range assumed for the entire phase will be too narrow if it is applied to the individual activities that comprise that phase.

Individuals often underestimate the range on the ten short activities because they do not appreciate the implication of a sample size of one activity. People do not appreciate that any one activity can have ranges that are remarkably wide. An interesting illustration of that fact can be found when we ask: 'What are the ranges applied to the ten individual activities that, when simulated, result in the 80 percent and 160 percent impact range for the path?' The ten individual activities would need ranges of 4.5 days–10 days–21 days to yield a phase that has a range of 80 days–100 days–160 days. In other words, instead of the 80 percent and 160 percent ranges on individual activities we would need ranges of 45 percent to 210 percent to get the path result expected. The result of those ranges on the total path risk is shown in Figure 5.5 as consistent with the initial assessment of 80 percent and 160 percent.

If interviewees are looking at the ten activities they often assign to each of the individual risks the same percentage ranges that they apply to the entire path. In the interviews about individual activities they typically reject any wider ranges as being too extreme. They do not appreciate that individual activities can have, indeed must have, wider ranges to get expected ranges for the path.

How can we correct their input for the sample-size bias? There is a correction factor available in most software that can be applied when we believe the ranges are narrower than they should be because of a bias.

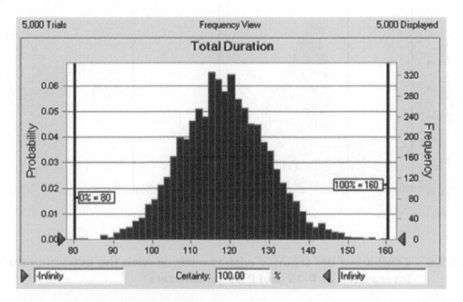

Figure 5.5   Result of path made up of ten activities with ranges of 4.5 Days, 10 days and 21 days

The Trigen Function

The triangular distribution, which is often used in schedule risk analysis, has a first cousin distribution, the 'Trigen' function. Short for 'triangle generation,' this Trigen function reinterprets the values given in the interviews and generates a new triangular distribution. This reinterpretation involves redefining the reported values as representing a range narrower than the extreme 0 percent and 100 percent that are used for the triangular distribution.

If the interviewer believes that the interviewee has provided values that are, say the 10 percent and 90 percent points of the triangular distribution, the Trigen function can be used. Suppose the interviewees have provided and insisted on values of 8 days and 16 days on the ten individual 10-day activities. We can apply the Trigen distribution to those values and make them represent the 15th percentile and the 85th percentile of a revised triangular distribution. As Figure 5.6 shows, this adjustment widens the 8 days optimistic to 16 days pessimistic impact range to a wider impact range that is 4 days optimistic and 21 days pessimistic. The comparison of the input values of the triangular and the Trigen functions is shown in Figure 5.6.

As we have seen previously in Figure 5.5, if ten individual activity impact ranges are adjusted in this way they provide the path risk that is equivalent to the desired 80 percent to 160 percent for the ten-activity path. The Trigen function adjustment corrects for narrow ranges on individual activities.[6]

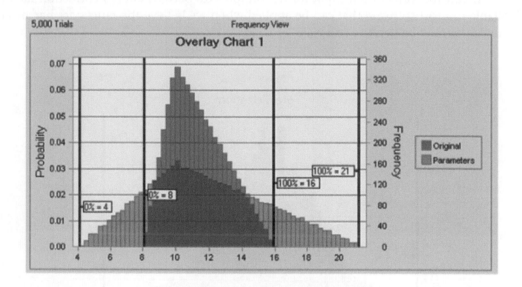

Figure 5.6   Comparing the triangle with the trigen with the extreme values interpreted as P-15 and P-85 parameters

Summary of risk data collection methods and problems

The quality of the risk data determines the success and credibility of the schedule risk analysis. Gathering risk data is the most important activity in a risk analysis exercise. Data gathering takes more time and requires more understanding and sophistication by the risk analyst than do the knowledge of risk analysis software or statistics.

Risk data are gathered from people in interviews or workshops. The risk analyst/interviewer must understand the dynamics of those environments, the pressures from the organization on the individual and the motivational and cognitive biases that often occur, particularly when interviewing people new to the practice of risk analysis.

The chapter covers the types of risk data needed. Risk probability and impact ranges are the most common data but other data such as correlation coefficients are needed as

---

6   The same correction factor also works well with the BetaPERT distribution in some simulation software.

well. We are interested in risk data that encompasses at least three types of uncertainty, estimating error, variability and uncertainty caused by discrete risk events.

Problems in collecting risk data come from the individual and from others in the organizational environment within which the people providing the data operate.

The corporate or organizational culture is often hostile to, or at best indifferent to, conducting project risk analysis. People who talk about project risk, even in workshops scheduled for that purpose, may be punished for offering their honest assessment of risk. We have discussed organizational risk management maturity which includes structures and practices that make the people comfortable and give them the tools for realistic risk assessments. We also recommend assuring interviewees of the confidentiality of their inputs, providing cover for them to provide honest and realistic risk data during interviews.

Individuals experience difficulty in providing risk data because the concepts may be new to them. They may have limited or no experience or training in providing risk data and lack access to databases than might help. Individuals providing data about uncertainty ranges and probability typically rely on heuristics or rules of thumb to calibrate their judgment. While heuristics can help people make useful estimates of risk concepts, some well-known biases exist that mostly lead to underestimating project risk parameters:

- The representativeness class of bias includes being insensitive to historical probabilities or experiences, which means ignoring the bulk of history when discussing a current project. Included here also is insensitivity to sample size, which in this context implies that people tend to use data that may be appropriate for many projects or for entire project phases to apply to individual activities on a specific project. It is difficult for interviewees to imagine the high probabilities and wide ranges that would be appropriate for specific activities. Also there is the illusion of validity or unwarranted confidence in the on-plan success of individual activities that occurs under this class of biases. This bias often leads to underestimating project risk.

- Availability bias leads people to estimate the risk of a project based on the past events that are most easily brought to mind. Those past events, which were both dramatic and personal to the interviewee, may bias interviewees' reporting of risk data. This bias can lead to over-estimating or under-estimating risk.

- A specific case of the availability bias might lead to an over-reaction to events that are dramatic and very current. Many people will react to the last information they are given and perhaps throw out earlier project assessments, however well considered. A rational (Bayesian) approach would be to combine the prior and current information correctly for a revised estimate of risk.

- Adjusting and anchoring leads to SMEs' providing risk impact ranges that are too narrow and probability of the risks that are too low for the risks they are describing. Their estimates are based on inadequate or small adjustments made from the anchor of values that appear in the baseline schedule. Narrow impact ranges and low risk event probabilities often occur during interviews from this source of bias. We have explored a useful way to adjust the risk ranges provided, using the Trigen function with the interviewee-provided ranges reinterpreted to represent user-defined percentiles, to offset this underestimation if it is perceived.

Risk interviews may be for the purpose of collecting detailed or high-level summary data, for confirmation of data by getting second opinions, for getting information from sources independent of the project or for background data. The interview itself is described.

The interview for people without experience in risk analysis usually includes an introduction to familiarize the interviewee with the process and answer questions. The introduction should also include a statement about the importance of risk analysis delivered by the project manager or someone else in authority. The facilitator asks the interviewee to describe pessimistic cases before calibrating them, since that is the most difficult value in the impact range for most people. The interview may start slowly if the interviewee is new to the process, but it will pick up speed as it goes along. However, interviews can take on average 1–3 hours, and may take even longer, depending on the interviewee and the amount of data required.

Interviews proceed more quickly if the interviewee is provided a format for the required information in advance and prepares for the interview by filling in the probability and impact range data in advance. Interviewees should expect challenges from the facilitator, however, particularly about low probabilities and narrow impact ranges. This is not because the facilitator wants to show more risk than there is, but rather because they recognize the common biases associated with the heuristics many people rely on to form their data.

The interviewee needs to understand the frame of reference for the risk analysis. The analysis is concerned with all risks that can affect the project. This includes technical, organizational, external and even project management risks, whether the organization can do anything about them or not. Whether change orders are included depends on the perspective of the client—the owner would consider change orders to be risks but the contractor will consider them as constituting new work for which additional money and time will be allotted. Some 'acts of God' such as hurricanes, earthquakes and fires are probably out of scope, although the effect of weather may be included.

Risk data are sometimes collected in a risk workshop. The workshop has some benefits in getting a number of people's views at once. There is synergy between the participants that can uncover new risks and considerations about risk calibration that is absent in one-on-one interviews. One problem with the workshop is the lack of anonymity and the possibility of pressure exerted on participants by their peers or supervisors to avoid certain types or risks or to shade the risk data away from serious implications for the project. In the interview, as distinct from the workshop environment, the facilitator can promise anonymity, which usually helps participants talk honestly about project risk.

Collecting realistic data about project risk is more challenging and important than learning about the Monte Carlo software. A successful risk analyst or facilitator will combine understanding of the possible biases with a way to interact with the interviewees to get good-quality risk data.

References

Barclay, S. (1977). *Handbook for Decision Analysis*. Defense Advanced Research Projects Agency.

Christensen, D. (1993). 'The Estimate of Completion Problem; A Preview of Three Studies.' *Project Management Journal*: 37–42.

Gilovich, T. et al. (2002). *Heuristics and Biases: The Psychology of Intuitive Judgment.* New York, Cambridge University Press.

Hall, D. et al. Risk Management Maturity Level Development. *Risk Management Research and Development Program Collaboration.*

Hillson, D. (1997). 'Towards a Risk Maturity Model.' *International Journal of Project & Business Risk Management* 1(1): 35–45.

PMI (2004). *A Guide to the Project Management Body of Knowledge.* Newtown Square, PA, Project Management Institute.

Salim, Y. and Hulett, D. (2008). Assessing Project Risk Management Maturity in a Large Energy Company. *Asia-Pacific PMI Congress.* Sydney, Australia, Project Management Institute.

Tversky, A. and Kahneman, D. (1974). 'Judgment under Uncertainty: Heuristics and Biases.' *Science* 185: 1124–31.

US AirForce (1973). 'Final Report,' US Air Force Risk Analysis Study Team.

USOMB (2006). Capital Programming Guide OMB.

Wikipedia (2008). Bayes' Theorem.

6 Where Parallel Paths Merge: Introducing the 'Merge Bias' and Risk Criticality

Introduction

Chapter 3 illustrated Monte Carlo simulation using as an example a simple one-path schedule. That example illustrated the method of analysis, but was not very realistic because very few project schedules have only one path. Most schedules have multiple paths. One path may be somewhat longer than the others and is identified as Critical in the sense of that term used in the Critical Path Method (CPM) of scheduling, but there are other Near-critical or Slack paths that may contribute to the schedule's risk.

Most schedulers and project managers will look at the CPM schedule and focus on the date that is given there, even if they do not entirely believe it. That date is determined by the critical path (there may be more than one, but there is at least one).

Slack paths are not as long as the critical path and therefore they have some flexibility or Float. In critical path methodology, a path with positive total float can be started later or elongated by the amount of its total float without delaying the final completion date of the project. This principle is true, of course, only if 'everything goes according to plan,' an assumption which is not common among projects.

The questions addressed in this chapter are: 'Why do we need to examine the risks on the paths that are not CPM-critical? Why is it not enough to address the risk on the critical path and focus less on the risk on the other paths?' The answer to these questions lie in the concept of Merge Bias. Merge Bias is the extra risk that occurs at points where parallel paths merge in the schedule. The risk at merge points may be greater than the risk of any of the merging paths.

As a side but curiously persistent issue, in an Appendix we examine the old Program Evaluation and Review Technique (PERT) and demonstrate that PERT does not give the correct answers for projects with parallel paths and merge points because it does not recognize the crucial merge bias.

The Importance of Off-Critical Paths at Merge Points

Experienced schedulers will be alert to the potential for mischief from paths with risk and positive but small total float. They realize that these paths may easily slip in a dynamic project and become critical, driving the completion to a later date. Most schedulers and project managers have experienced projects where the path that ultimately drives project completion is not the path that was initially identified as being critical. The critical path may change during the project's execution. For this reason some project managers define critical as any path with total float of less than some positive number of days, for example 30 days, rather than just zero days. This consideration is essentially recognition that, given the uncertainty of the project, off-critical paths can come into play and determine the completion date.

The essence of schedule risk is that the duration of every project path is uncertain. The uncertainty of individual activities along each path is combined according to the logic of the schedule into total path risk as shown in Chapter 3. To reiterate the results of that chapter, the duration of a single path schedule may be longer or, sometimes, shorter than originally indicated by the deterministic activity durations. The project duration (and completion date) will be determined by the duration of that path. If there is only one path in the schedule, that path will be *de facto* critical. The project will finish earlier if that path duration happens to be shorter, for instance if it is below its scheduled duration.

Something quite different happens when there are several project paths, however. It is no longer always true that if the critical path is shorter then the project is shorter. This is because of schedule logic at merge points. At a merge point, both (or all, if there are more than two) merging paths must complete for the project to complete. In Figure 6.1, both Unit 1 and Unit 2 must complete for the finish milestone to be considered complete.

Put another way, either Unit 1 or Unit 2 (or both) can delay the project. The project can finish early only if both Unit 1 and Unit 2 finish early. Because of this fact, a path that is not critical if everything goes according to plan may become critical and delay the project as durations change during execution. The slack path may be important in actuality, even though it does not appear to drive the project if the deterministic durations are accurate, because slack paths may turn out to be longer than critical paths. This is a problem that is explored by schedule risk analysis.

Real project schedules have many merge points, and some of these have multiple merging paths. In the project plan there are many events where two or more paths have to be complete. One such event may be the beginning of the Integration and Test phase where two or more subsystems must be finished so we can integrate them into a system for testing. Other merge points may be decision-points or stage-gate reviews where several estimates, reports and analyses are inputs necessary before the decision can be

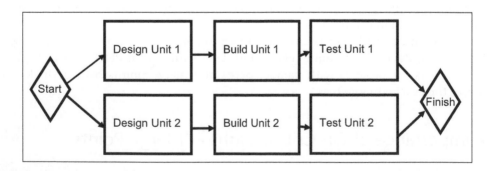

Figure 6.1   Simple two-path schedule

made and the project can proceed. For instance, the beginning of product assembly often requires components that arrive on their own schedules and have to be coordinated before assembly can begin.[1] At points such as these, it is no longer true that: 'The project will finish earlier if that path duration happens to be shorter,' as it is with a single-path schedule (see Chapter 3). At the merge points all paths need to be shorter for that event to occur earlier, and it is easy to imagine circumstances under which the event will be delayed.

Suppose the following simple example:

* The path for Unit 1 is expected to take 100 days to get to the point where it can be integrated. Its uncertainty is between 85 and 130 days, optimistic to pessimistic duration.
* The path for Unit 2 is expected to take 95 days. It too has a range, say from 80 days to 125 days.

With a single subsystem, say Unit 1, there is a chance that it alone could take 90 days and the integration phase could start 10 days earlier. The complication arises when Unit 2 is also required for Integration. If Unit 2 is longer, say 112 days, it does not matter that Unit 1 completes in less than 100 days, the project is 12 days late at that point because of Unit 2, which was not critical in the CPM sense.

There is no assurance in the typical project that Unit 1 and Unit 2 will both complete on time. Unless there are some reasons why their durations move together (we say they are 'correlated,' see Chapter 10), one path can be long while the other path is long, short or close to the duration as scheduled. The determination of when a milestone event occurs is: 'The latest early date of the merging paths.' Hence it is easy to see that either Unit 1 or Unit 2 could make this simple merge point late by itself, irrespective of the finish of the other path. Of course, if both paths are late the merge point will be late. The only chance for the merge path to be early is if both (or all, if there are more than two) merging paths are early.

Review of the Risk in the Single-Path (Unrealistically-Simple) Project Schedule

The simple project shown below has one path. Reviewing the methods and concepts presented in Chapter 3 we can find the risk to the finish milestone. Figure 6.2 shows the one-path project we will use.[2]

Let us interview some experts in design, building, testing and shipping. Suppose we find out that, in this project, the most likely durations are those deterministic durations in the schedule but there are optimistic and pessimistic durations that should be considered. The results of these interviews are shown in Table 6.1.

---

1   In one project an assembly point had over 15 merging paths, many of them at risk for not making the assembly start date. The company was forever scrambling at that point to get the needed components in time.

2   We show results from several software packages. This one is generated in Pertmaster® from Primavera Systems.

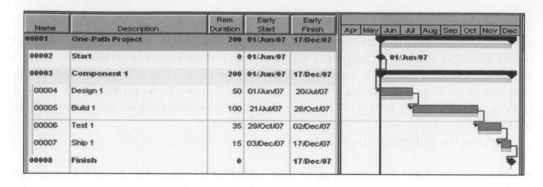

Figure 6.2   One-path project schedule

Table 6.1   Example of risk ranges collected during risk interviews

Data Derived from the Interviews				
Activity	Duration	Optimistic	Most Likely	Pessimistic
Single-Path Project	200			
Start	0			
Design	50	40	50	75
Build	100	80	100	140
Test	35	25	35	70
Ship	15	10	15	20
Finish	0			

Without including risk, this project finishes on December 17. We are committed to that date, but we might want to consider the impact of project risk on three important questions:

1. What is the likelihood that this project will finish on December 17 or sooner?
2. What is the average date this project would finish on if it were executed many times with the risks listed in the table above?
3. If we would like, say, an 80 percent confident date, one where there is only a 20 percent chance of overrunning, is that December 17 or some other, possibly later, date?

Repeating the basic approach of Chapter 3, Monte Carlo simulation of the single-path schedule provides the schedule shown in Figure 6.3.

Simulating the original single-path schedule 5000 times gives a result as shown in Figure 6.4.

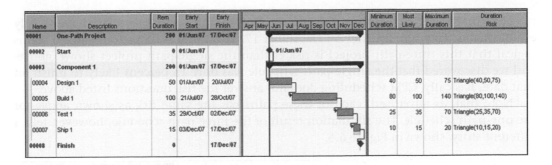

Figure 6.3   Schedule with three-point estimates inserted

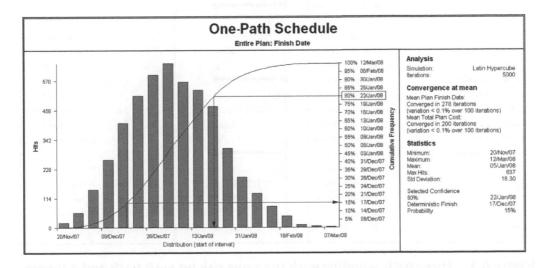

Figure 6.4   Results of a Monte Carlo simulation of the simple single-path schedule

- The schedule date, December 17, is only about 15 percent likely to be achieved, so we are 85 percent likely to overrun.
- The average result is January 5 with the standard deviation of 18.3 days. If we conducted this project many times, on average we will be about 3 weeks late.
- If we want a date that is 80 percent certain to be achieved we should adopt January 22.

Example of a Three-Path Schedule—The Merge Bias at Work

Let us examine an example of parallel paths and a merge point, a schedule structure that would be very common in project plans. We will encounter the merge bias. The schedule we will use has three paths that merge at the project finish milestone. Let us assume that the risk on each path is the same as in the single path schedule above. This new schedule is shown in Figure 6.5. In our example we have assumed that each component has the same four activities and the same risks on each.

Note that in CPM scheduling, this project, like the single-path project we have already established is risky, also finishes on December 17. This result is suspicious since it seems logical that this three-path project is riskier than the single-path project shown above, and we discovered that the single-path schedule was only 15 percent likely to finish on that date. Basically, CPM scheduling does not answer the risk questions listed above.

Note that, as advertised, each of these paths has the same risk as shown above for the one-path schedule. The simulation result of the three-path schedule, however, tells a different story, shown in Figure 6.6.

Name	Description	Rem Duration	Early Start	Early Finish	A M J J A S O N D	2008 J F M	Minimum Duration	Most Likely	Maximum Duration	Duration Risk
00001	Three-Path Project	200	01/Jun/07	17/Dec/07						
00002	Start	0	01/Jun/07		01/Jun/07					
00003	Component 1	200	01/Jun/07	17/Dec/07						
00004	Design 1	50	01/Jun/07	20/Jul/07			40	50	75	Triangle(40,50,75)
00005	Build 1	100	21/Jul/07	28/Oct/07			80	100	140	Triangle(80,100,140)
00006	Test 1	35	29/Oct/07	02/Dec/07			25	35	70	Triangle(25,35,70)
00007	Ship 1	15	03/Dec/07	17/Dec/07			10	15	20	Triangle(10,15,20)
00008	Component 2	200	01/Jun/07	17/Dec/07						
00009	Design 2	50	01/Jun/07	20/Jul/07			40	50	75	Triangle(40,50,75)
00010	Build 2	100	21/Jul/07	28/Oct/07			80	100	140	Triangle(80,100,140)
00011	Test 2	35	29/Oct/07	02/Dec/07			25	35	70	Triangle(25,35,70)
00012	Ship 2	15	03/Dec/07	17/Dec/07			10	15	20	Triangle(10,15,20)
00013	Component 3	200	01/Jun/07	17/Dec/07						
00014	Design 3	50	01/Jun/07	20/Jul/07			40	50	75	Triangle(40,50,75)
00015	Build 3	100	21/Jul/07	28/Oct/07			80	100	140	Triangle(80,100,140)
00016	Test 3	35	29/Oct/07	02/Dec/07			25	35	70	Triangle(25,35,70)
00017	Ship 3	15	03/Dec/07	17/Dec/07			10	15	20	Triangle(10,15,20)
00018	Finish	0		17/Dec/07						

Figure 6.5   Three-path schedule with the same risk on each path and a merge point

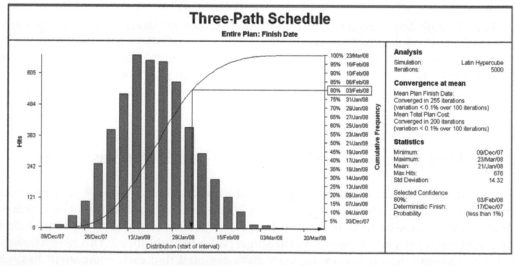

Figure 6.6   Monte Carlo simulation results for the three-path schedule

The results for the Monte Carlo simulation shown in Figure 6.6 are different for the three-path schedule versus the single-path schedule:

- With three merging paths, the average completion date is January 21, not January 5. On average we are 5 weeks, not 3 weeks, late.
- The 80th percentile is February 3, not January 22, when you have three identical paths. For this schedule we need a contingency reserve of 7 weeks, not 5 weeks, to provide the degree of certainty (80 percent) we desire.
- The probability of hitting December 17 is less than 1 percent, not 15 percent as it is for each of the paths considered individually.

The cumulative distributions in Figure 6.7 compare the results for the single-path schedule with that for the three-path schedule. In our example each path in the three-path schedule has the one-path cumulative distribution, or S-curve, but the schedule that includes three paths merging at a common finish milestone has the three-path S-curve.

The effect of the merge bias is shown by the shift from the single-path schedule's cumulative distribution to that of the three-path schedule, as highlighted by the horizontal arrow in Figure 6.7.

The merge bias has been known since the early 1960s through the work of two researchers at the Rand Corporation, MacCrimmon and Ryavek, (1962) which was included in an appendix of a leading book on CPM scheduling by Archibald and Villoria (1967) in the mid-1960s.

This added risk at merge points occurs, potentially, at each merge point in the schedule. This is one reason why complex projects with many parallel paths and merge points are potentially riskier than simpler projects with fewer paths. It explains why project schedules tend to overrun more than just the duration risk would indicate, and why some merge points are viewed by project managers as sources of high risk.

The logic of the merge bias occurs, potentially, not just at the finish milestone, but at any merge point in the schedule.

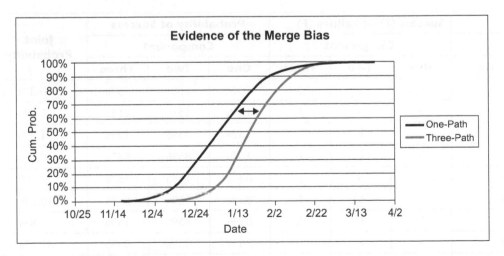

Figure 6.7   Graphical evidence of the merge bias from Monte Carlo simulation cumulative distributions

The Merge Bias is a System Problem

There is one circumstance when a merging path does not contribute to the risk at the merge point, however. A merging path may have so much float compared to its inherent risk that it is not likely to be an issue at a merge point. For instance, consider the case where a merging path has 100 days of float to the merge point but only 50 days of downside (threat) risk at the maximum. Planning the project to start the slack path as soon as possible, using an 'early starts' strategy, should effectively keep it from contributing to the risk at that merge point.

Of course, there is a potential problem with this scenario lurking within the deterministic (durations are known with certainty) CPM system of analysis. CPM scheduling tells us that a path with 100 days of float can be started on day 101 (this is called 'late starts') and still not delay the project. Based on the discussion of this chapter we now know that starting a risky path that has 100 days of total float on day 101, as CPM scheduling says is possible without jeopardizing the project's completion date, makes that path critical to the merge point again, putting us back in the merge bias soup that is discussed above. Planning to start activities at their earliest date turns out to be a good plan, given the risk and merge bias.

What's going on here? Why would just including more paths merging cause extra risk? The reason is that almost any merging path can delay the project. In our example all three paths have to be on time or early for the finish milestone to complete on December 17. What is the likelihood of all three of the merging paths finishing on time or earlier? Each has the same risk in our example, so each has a 15 percent chance of finishing on December 17 or sooner. Table 6.2 shows the stark reality of what is happening at this finish milestone if each of the three paths is independent of the others.

Table 6.2   Eight possible scenarios—one indicating success meeting merge point

Case	Success (S) or Failure (F) Component			Probability of Success Component			Joint Probability
	One	Two	Three	One	Two	Three	
1	S	S	S	15%	15%	15%	0.3%
2	S	S	F	15%	15%	85%	1.9%
3	S	F	S	15%	85%	15%	1.9%
4	S	F	F	15%	85%	85%	10.8%
5	F	S	S	85%	15%	15%	1.9%
6	F	S	F	85%	15%	85%	10.8%
7	F	F	S	85%	85%	15%	10.8%
8	F	F	F	85%	85%	85%	61.4%
Sum							100.0%

The only case out of the eight logical cases in which the project finishes on or before December 17 is Case 1, where all three paths to the merge point finish on or before that date. With three paths each having a 15 percent chance of success, the chance of success at the merge point is only a 3/10 of one percent. Cases 2, 3 and 5 indicate that only one component may cause a late finish. Cases 4, 6 and 7 indicate that combinations of two components together may cause the project to finish late. In Case 8 all three components are late. In each of these seven cases, however, the schedule is late at the merge point. Hence there is a chance of 99.7 percent that this project will finish later than January 17.

The effect of the merge bias is the effect of putting together various components of a project into a system known as a schedule. The rule at the merge point milestone, that any path can delay the project but that all paths must be successful for the project to finish early, is asymmetrical but real. We need all components before we can test the integrated system, make a wise decision or assemble a product. This phenomenon is a problem only when the paths are included in a system of paths. Realistically, most projects have many parallel paths and many merge milestones, so this is the rule rather than the exception. Monte Carlo simulation handles the prospect that any one of the merging paths can delay the merge event.

Which Activities and Path will Ultimately Delay the Project?

Project managers need to conduct effective risk responses that increase the probability of schedule success. To do this they need to identify and focus on those activities and paths that determine how long the project will take. In traditional project management the critical path is the path that is the longest through the schedule and thus determines the earliest finish date of the project. Most project managers will identify and focus their risk mitigation efforts on the activities on the critical path, although they will probably identify 'critical' as those activities with low float rather than with no float.[3]

The problem with basing our responses on the CPM schedule's critical path is that, since we do not know the durations of activities with certainty, we do not know the critical path with certainty. Many schedulers and project managers brought up with the CPM scheduling mentality will argue with this statement since they do not understand that that CPM critical path may not be the critical path when the project finishes. Yet, these same experienced project managers know that the critical path may change during the project execution.

Why might the critical path change during project execution? Some risks may occur and cause a path that originally had some float to become longer than other parallel paths and therefore become critical. A common example of this phenomenon is a path that includes the issuance of an environmental or land-use permit. The original plan may assume that the governing agency will issue the permit according to its own procedures, say after a review period of 30 days from submittal. It is not uncommon, however, for the agency to require more time for the permit review or to have questions that require more study, submittals and even public hearings. If the permit were on a path with, say,

---

3   In a 3-year project a level of total float of, say, 30 days is not sufficient to keep the path from delaying the project. Much can happen in 36 months and a 1 month buffer (about 3 percent of the total schedule) is usually not significant.

20 days of total float, any delay of this magnitude could use up the float and the slack path including the permit activity may become critical.

Project schedule risk analysis can help the project manager identify in advance those risks and paths that look 'safe' in the current plan but which, under some circumstances, may become critical. These paths that originally have total float have enough risk potentially to delay the project.

We care about the changing critical path because the project manager needs to design and implement risk mitigation activities on the paths that will ultimately delay the project, even if they are not critical on the original schedule. Since risk analysis looks forward it is possible to predict which paths may become critical in the future, giving the project manager a 'heads up' warning so mitigation can focus on the right risks.

In most cases, even schedule risk analysis cannot identify which path will be critical. We can use the methods shown below to identify at least which activities and paths are most likely to delay the project.

A slack path occurs when its activities, with their durations and schedule logic, take less time than do those on a path that is parallel and takes longer. These paths might, if they are risky, become critical. Examine the three-path schedule shown in Figure 6.8. Its schedule began as the same schedule shown above in Figure 6.5. In this example, however, the paths for Components 1 and 3 are shorter (Components 1 and 3 are now slack paths) and Component 2 is identified as the critical path.

Once the critical path is identified in the base plan, we observe that the traditional project manager has successfully mitigated those risks and hence the risk ranges on the activities in Component 2 are narrower than those of the other two components. This project manager has focused only on the critical path at the expense of any other paths.

The paradox of this approach, managing risk only on the critical path and possibly ignoring slack paths, is that after successful risk mitigation the critical path is still critical but is in fact the path least likely to delay this project. As we see below, given the risks and the parallel path logic, Components 1 and 2 are more likely to delay this project.

ID	Description	Rem Duration	Start	Finish	May	Jun	Jul	Aug	Sep	Oct	Nov	Dec	2008 Jan	Feb	Minimum Duration	Most Likely	Maximum Duration
000001	Three-Path Project	200	01/Jun/07	17/Dec/07								▼					
000002	Start	0	*01/Jun/07		◊01/Jun/07												
000003	Component 1	195	01/Jun/07	12/Dec/07								▼					
000004	Design 1	45	01/Jun/07	15/Jul/07											35	45	70
000005	Build 1	100	16/Jul/07	23/Oct/07											80	100	140
000006	Test 1	35	24/Oct/07	27/Nov/07											25	35	70
000007	Ship 1	15	28/Nov/07	12/Dec/07											10	15	20
000008	Component 2	200	01/Jun/07	17/Dec/07								▼					
000009	Design 2	50	01/Jun/07	20/Jul/07											45	50	60
000010	Build 2	100	21/Jul/07	28/Oct/07											90	100	120
000011	Test 2	35	29/Oct/07	02/Dec/07											30	35	45
000012	Ship 2	15	03/Dec/07	17/Dec/07											12	15	18
000013	Component 3	190	01/Jun/07	07/Dec/07								▼					
000014	Design 3	50	01/Jun/07	20/Jul/07											40	50	75
000015	Build 3	90	21/Jul/07	18/Oct/07											70	90	130
000016	Test 3	35	19/Oct/07	22/Nov/07											25	35	70
000017	Ship 3	15	23/Nov/07	07/Dec/07											10	15	20
000018	Finish	0		17/Dec/07								▼ 17/Dec/07					

Figure 6.8   Three-path schedule with critical Component 2 successfully risk managed

We use several measures to determine which paths are most important to receive risk management attention and risk response resources, based on a Monte Carlo simulation.[4] Two of these measures are:

1. Risk criticality is the percentage of iterations that the activities are on the critical path. If an activity is critical in 700 of 1000 iterations in a simulation we say that it is 70 percent likely to be the path that will ultimately delay the project.
2. Risk sensitivity reflects the correlation between the activity duration and the duration of the total project schedule during simulation. A strong correlation is intended to indicate strong causality since we know that the individual schedule activities drive the schedule completion date, not the other way around.

Looking at risk criticality first, we see that, because of the successful risk mitigation, Component 2 is the least likely of the three components to delay this project. Component 1 is 47 percent likely and Component 3 is 36 percent likely to delay the project. The traditional CPM critical path, Component 2, is so well risk mitigated that it is only 19 percent likely to be the ultimate critical path. These results are displayed in Figure 6.9.

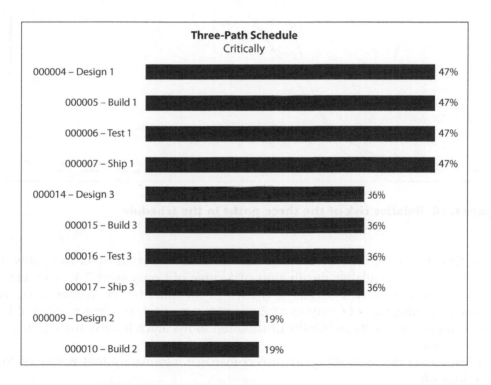

Figure 6.9   Risk criticality with successful mitigation of the critical path risk

4   Another measure of importance is risk cruciality. As used by Pertmaster, cruciality multiplies the risk criticality and sensitivity components for each activity. This computation intends to highlight those activities that are mostly on the critical path and are strongly correlated with, and hence drive, the completion date. This is a different definition from the one that Terry Williams first introduced in the early 1990s and he does not support this one (private communication February 4, 2008).

A project manager seeing these results might be pleased at the success of the risk mitigation applied to Component 2 to date. The focus of risk mitigation efforts going forward should be on the activities of Components 1 and 3, however, even though they are still not critical in the traditional CPM sense. These results show that not knowing the durations with certainty should lead to an investigation of risk criticality that is appropriate to an uncertain world where we can say what is most likely to delay the project though not usually what will delay the project.

Why is the critical path so unlikely to delay this project? The reason can be seen by looking at Figure 6.10 in which the cumulative probability distributions of the three paths are presented in an overlay chart.

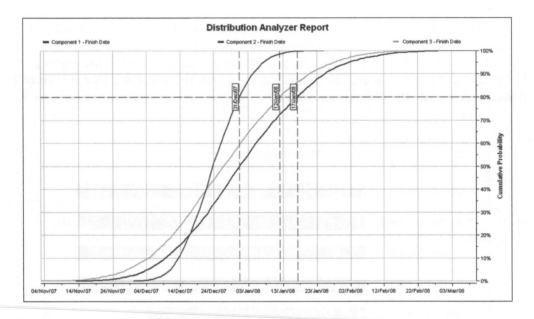

Figure 6.10 Relative risk of the three paths in the schedule

In this chart the risk of Component 2 is less than that of the other two paths. This is shown by the cumulative distribution or S-curve of Component 2 that is standing straighter and crosses the P-80 line to the left of the other two components' S-curves. Because risk mitigation of Component 2 was very successful, and hence reflected in its narrow impact ranges, its probability distribution shows much less risk than those of the other components.

Clearly criticality is showing the project manager information that will be useful in risk mitigaton:

• Looking only at the path durations in the schedule can mislead the project manager about the locus of the project risk and hence where risk management attention should next be placed.

- Project risk analysis using Monte Carlo simulations can provide a much more relevant measure of which path is risky by showing which path is most likely to be the path that ultimately delays the project.
- Risk criticality can be calculated early in the project and must be calculated periodically as unknown risks emerge, known risks change or are retired and as risk responses implemented are successful or not.

Another measure of importance or priority of activities is duration or schedule sensitivity, which is shown in Figure 6.11 for the current example schedule. These measures reflect the correlation between the activity duration and the project schedule duration during simulation. Duration sensitivity reflects both the risk in the activities and their relationships in schedule logic. While correlation between two uncertain elements does not usually imply causality, in this case the activity duration risks do cause overall schedule duration risk. This measure can be used to indicate which activities are both likely to delay the project and are highly variable.

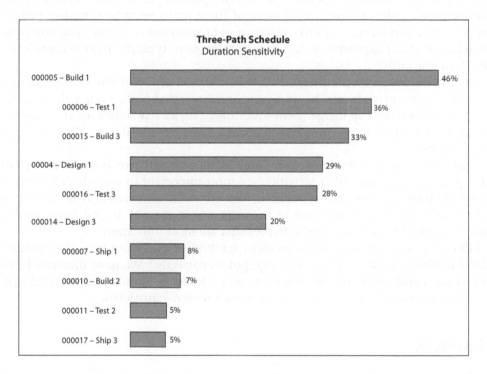

Figure 6.11 Duration sensitivity between activities and risk of the total schedule

Using duration sensitivity, we see that Build Component 1 is the most important activity driving the project schedule, followed by Test Component 1 and Build Component 3 as important activities driving the schedule's duration. These results are consistent with the results in Figure 6.9 concerning risk criticality. Criticality, however, only identifies the path but not the activity driving that path to be important. These two measures together can provide an indication for the next most important risk mitigation actions.

It is possible, of course, for an activity that is risky but not important to be highly correlated with the overall schedule duration. This may occur, for example, if the activities' durations are correlated (see Chapter 10)[5] then the measures of their duration sensitivities will be similar. If one of these activities shows a strong correlation to the schedule's duration, the other one will as well, even if it is never on the critical path or its duration is small and insignificant and cannot drive the schedule.

Summary: Real Schedules have Parallel Paths and Risk Builds Up at the Merge Points

In Chapter 3 we introduced the fundamental analytical tool, Monte Carlo simulation, by using it on a simple single-path schedule. In this chapter we have made the schedule more realistic by introducing parallel paths. With parallel paths there will be path merge or convergence points where two or more of these paths must be complete before the next project step is taken. Steps like integration of subsystems, assembly of components or making of major expenditure decisions at stage gates typically require completion of more than one path.

The phenomenon of the merge bias is introduced at this point and we show that risk builds up at merge points because any single merging path, even a slack path, has the potential to delay the merge point milestone. Project risk builds up at merge points throughout the project and experienced project managers are alert to all merging paths for this reason.

Once we consider schedules with more than one path we become interested in which path, and indeed which activities, are most important in the schedule's risk. Two commonly used measures are introduced, risk criticality and duration sensitivity. Each of these tells the analyst and project manager where the risk is and helps guide risk responses toward those paths and activities that are most important.[6]

Monte Carlo simulation is the modern method of analyzing project schedule risk and it handles the risk at schedule merge points correctly.[7] We have illustrated that in several charts and tables. Happily there are many schedule simulation packages that are accessible, inexpensive and easy to use even for the non-statistician.

References

Archibald, R. D. and Villoria, R. L. (1967). *Network-Based Management Systems*. New York, John Wiley & Sons, Inc.

---

5   Two activities' durations are correlated usually because they are driven by the same risks (see the risk driver approach described in Chapter 8).

6   In Chapter 8 we will introduce the risk driver approach that focuses on which risks (not which paths or activities) are most important to manage.

7   In the Appendix the failure of the Program Evaluation and Review Technique (PERT) to capture the extra risk at merge points is discussed. That is why PERT should not be used as a substitute for simulation in conducting a quantitative schedule risk analysis.

MacCrimmon, K. and Ryavek, C. (1962). An Analytical Study of the PERT Assumptions. *Research Memoranda*, The Rand Corporation. RM-3408-PR.

Williams, T. M. (1992). 'Criticality in Stochastic Networks.' *Journal of the Operational Research Society* 43: 353–357.

Williams, T. M. (1993). 'What is Critical?' *International Journal of Project Management* 11: 197–200.

7 Probabilistic Branching: Analyzing Discrete Risk Events

Introduction

Until now we have represented project schedule risk as uncertainty in the duration of individual tasks that have to be performed. Project schedule risk analysis recognizes that estimates of duration are not certain. Durations that are typically viewed as single-point or deterministic values are better represented by probability distributions than by single values such as '30 days.'

We have argued that these distribution ranges—including the optimistic, most likely and pessimistic durations—can be determined with some accuracy, in part by examining existing data and recently completed similar projects, but mostly by in-depth interviews of people involved with and knowledgeable about the project. The type of duration uncertainty has been considered as a continuous distribution with a 100 percent probability of occurring. We call that 'uncertainty' because it represents an uncertain estimate of duration.

In this chapter we turn to the representation of risk events and how they can be represented in project schedule risk analysis. A hallmark of a risk event is that it may or may not occur on your project. If it occurs it impacts some aspect of the schedule to, generally, an uncertain extent. The representation of the uncertainty of occurrence of an event-type risk in Monte Carlo simulation is in the number or percentage of iterations in which the risk appears.

Adding risk events that may or may not occur agrees with a well-known definition of project risk: 'An uncertain event or condition that, if it occurs has a positive or negative impact on at least one project objective.' (PMI 2004).

Risk Represented by Uncertain Risk Events

This chapter focuses on a different type of project activity risk, risk events that may or may not occur. If they do occur, these risk events may cause work that is not typically found in the project schedule. For instance, a system component may fail a crucial performance test. If that happens there is a series of activities including:

- finding the root cause of the failure;
- deciding what to do to remedy that failure;

- doing the chosen activities; and
- retesting the modified component.

These four activities do not need to be included in the schedule if the component passes the test, and hence they are not usually included in the schedule at all. Indeed, most schedules are 'success oriented' and do not include recognition of the possibility of failure or serious problems. This may be because project managers want to drive their team to success, even though they know that serious problems may de-rail the project. Also, recognizing risk may be viewed as self-fulfilling, since 'SAISS' (schedule allocated is schedule spent), a schedule-analogue to the more common 'MAIMS' (money allocated is money spent).

And, yet, we would be foolish and it would be unrealistic to assume that tests cannot be failed: in fact, tests are performed precisely because they may fail. That is why the tests are required for acceptance of the component or system. Reality is what a schedule risk analysis represents.

This chapter examines the treatment in project risk analysis of the class of risks that is sometimes called 'risk events.' The exciting characteristic of these risks is that they have a probability of less than 100 percent of existing and therefore are either included in the schedule or not. In practice, the schedule risk analyst will have to create new activities in the schedule to represent these risk events.

Relationship to Qualitative Risk Analysis: Analyze High-Priority Risks to the Schedule

Probabilistic activities are similar to the risks identified in the qualitative risk analysis process that leads to the commonly found risk register (PMI 2004). Qualitative risk analysis takes identified risks and assesses their:

- probability of occurring on this project; and
- if they were to occur, their impact on project objectives such as time, cost, scope and quality.

The probability of a risk's occurring and its impact if it were to occur are assessed by interviewing people knowledgeable in the project. Risks can be prioritized into at least three main groups: high (red) risk, moderate (yellow) risks and low (green) risks to a specific objective such as the schedule. Prioritizing risks in this fashion is a main result of the Probability and Impact (P×I) Matrix. The input to this approach is to define clearly what the stakeholders view as being (1) an impact and (2) a probability that would be considered to be anywhere from very low to very high for each objective, then use these definitions in the evaluation of each identified risk.

That P×I matrix also contains the stakeholders' view of the combinations of probability and impact that make a project risk worthy of special attention, perhaps for further study or direct handling (mitigation of a threat is the most common example of a response to a project risk). An example of a P×I matrix, including opportunities and threats, focused on the schedule is shown in Figure 7.1.

Probability and Impact Matrix for the Time Objective										
Probability	**Threats**					**Opportunities**				**Probability**
Very High	L	M	H	H	H	H	H	H	M	L	Very High
High	L	M	M	H	H	H	H	M	M	L	High
Mod	L	L	M	H	H	H	H	M	L	L	Mod
Low	L	L	M	M	H	H	M	M	L	L	Low
Very Low	L	L	L	M	H	H	M	L	L	L	Very Low
	Very Low	Low	Mod-erate	High	Very High	Very High	High	Mod-erate	Low	Very Low	
	Impact					Impact					

Figure 7.1 Probability and impact matrix for threats and opportunities—time objective

The result of the P×I matrix approach, or any other approach that prioritizes risks, is a key result shown in the risk register. The risk register is an important document of risk management that also includes identified risks, scored or rated according to their importance, and then the handling actions that are agreed upon.

The most common way to use this approach in quantitative schedule risk analysis is to focus on those risks that fall into the red or high risk (designated in Figure 7.1 with a 'H') for the time objective.

In quantitative schedule risk analysis, it is common to refer specifically to those identifiable risks that are judged to be high risk to the schedule during the risk interview. We are prepared to create a 'probabilistic branch' based on the information about risks' probability of occurring and impact on the schedule. Using the test success-failure example, the probability of failing the test and the duration needed if it were to happen (to accomplish the root cause analysis, the planning of remedial actions, the executing of those plans and the retesting of the article) would be typical entries in the risk register. That is accomplished by the probabilistic branch capability that is included in schedule risk analysis Monte Carlo simulation software.

These risk register risk events are:

- discrete because they will either exist or not with some specific probability; and
- discontinuous from the underlying risk. That means that recovery activities performed after a test has failed cannot be contained in a long-tailed distribution of the Test activity duration itself. Recovery from a failed test is distinct from continued testing.

Because of these two factors, specific risk register risks have to be handled differently from other risks in the schedule.

A Simple Example of Probabilistic Branching: Can We Fail the Acceptance Test?

Suppose we have a really simple one-path project. The tasks are Design, Build, Test and Ship. There is only one path through the schedule and activities are related by finish-to-start logic. In a previous chapter we introduced the notion of the uncertainty surrounding the estimates of duration for each of these activities. We also introduced the tool of Monte Carlo simulation, which is the accepted approach to combining the distributions along the path to reach conclusions about when the project will finish with a range of degrees of certainty. We use these methods on probabilistic branches introduced in this chapter as well.

The one-path project we will use is shown in Figure 7.2.

Let us interview some experts in design, building, testing and shipping. Suppose we find out that the most likely durations are those deterministic durations in the schedule but there are optimistic and pessimistic durations that should be considered. The results of these interviews are shown in Table 7.1.

ID	Description	Rem Duration	Start	Finish	2009 May Jun Jul Aug Sep Oct Nov Dec Jan Feb
One Pat	One-Path Schedule	143	02/Jun/08	17/Dec/08	
A1000	Start Milestone	0	'01/Jun/08		'01/Jun/08
A1010	Design	50	01/Jun/08	20/Jul/08	
A1020	Build	100	21/Jul/08	28/Oct/08	
A1030	Test	35	29/Oct/08	02/Dec/08	
A1040	Ship	15	03/Dec/08	17/Dec/08	
A1050	Finish Milestone	0		17/Dec/08	17/Dec/08

Figure 7.2   One-path project schedule[1]

Table 7.1   Example of risk ranges collected during risk interviews

Risk Ranges from the Interviews				
Activity	Duration	Optimistic	Most Likely	Pessimistic
Single-Path Project	200			
Start	0			
Design	50	40	50	75
Build	100	80	100	140
Test	35	25	35	70
Ship	15	10	15	20
Finish	0			

---

1   This figure uses Pertmaster® v. 8.0 from Primavera.

Without including risk, this project finishes on December 17. We are committed to that date, but we might want to consider the impact of project risk on three important questions:

- What is the likelihood that this project will finish on December 17 or sooner? The answer from the simulation, shown in Figure 7.3, is that there is only a 15 percent likelihood that the project will finish on or before December 17.
- What is the average date this project would finish on if it were executed many times with the risks listed in the table above? If this project were to complete 'on average,' the date would be January 6, not December 17. (With a 7-day calendar the project personnel have to work through the holiday season even to accomplish January 6.)
- If we would like, say, an 80 percent confident date, one where there is only a 20 percent chance of overrunning, is that December 17 or some other, possibly later, date? The 80th percentile is January 22, so we need a schedule contingency of (January 22–December 17) 36 days to achieve a conservative 80th percentile level of certainty. The contingency reserve of 36 days is 18 percent of the baseline duration of 200 days.

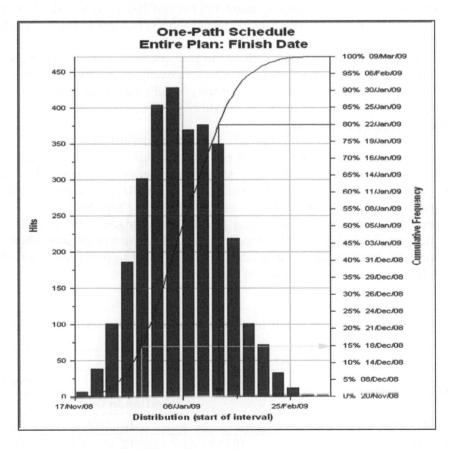

Figure 7.3   Results of a Monte Carlo simulation of the simple single-path schedule

Adding a Discrete Risk Event: Failing the Test

IMPACT SHOULD THE RISK EVENT OCCUR

Suppose that a risk that we may fail the test arises during the risk interviews. The subject matter experts (SMEs) assess that we may fail the test with a 30 percent probability. Those same SMEs then specify the scenario associated with failing the test, were it to occur. They develop the following table (Table 7.2) of events that would follow a test failure.

These activities are not included in the project manager's original schedule. In fact, few schedules are explicit about the possibility of failing the test, failing to get a permit, experiencing a labor strike or delay in management decision making. Most project managers will avoid recognizing the possibility of failure because:

- they want to push the team to succeed;
- it is embarrassing to admit that failing the test is a possibility.

However, while these are each events to be avoided or protected against, they may happen on your project. In fact, the scenario 'fail the test' risk ranges should be evaluated with full knowledge of the various risk mitigation actions that are in place at the time of the interview.

Since the activities are not often included in the schedule, we need to put them in as new activities. However, typical scheduling software assumes that activities in the schedule are part of the plan—that they are needed for the project's successful completion. Typical scheduling software will incorporate the durations of the recovery activities as they are inserted in the appropriate paths.

Test failure, at least of the integrated article, is often on a critical path since it is a final requirement before shipping the article. A process plant could experience such a failure during the commissioning phase with the same result. Hence, it is very likely that if the activities were inserted in the schedule with their 'most likely durations,' if they exist they would delay shipment of the deliverable in the baseline schedule.

Yet from the interviews we know that test failure is judged to be only 30 percent, not 100 percent, likely. We need to recognize that fairly substantial probability during the schedule risk analysis. At the same time we do not want to penalize the schedule for activities that are 70 percent likely not to occur.

Table 7.2   Risk ranges for activities that occur only if the article fails the test

| Probabilistic Activity | Scenario: Fail the Test | | |
| | Optimistic | Most Likely | Pessimistic |
	Duration in Days		
Root cause analysis	5	10	20
Determine the recovery plan	10	12	25
Execute the recovery plan	10	15	25
Retest the repaired item	15	20	35

The simplest and most easily understood solution is to include the activities in the schedule with zero days (0 days) duration so they do not affect the current schedule. We do need to include the activities' risk ranges, if they do occur, in the risk database.

In Figure 7.4, notice that the schedule duration is '0d' while the three-point estimates of duration appear in the OPTD, MOST and PESS columns as with the other risky activities. Since the schedule duration is 0 days the deterministic schedule still finishes on December 17.

Activity ID	Activity Name	Original Duration	Start	Finish	OPTD	MOST	PESS										
									Jun	Jul	Aug	Sep	Oct	Nov	Dec	Jan	Feb
☐ One Path Project One-Pat...		143d	01-Jun-08	17-Dec-08	195	257	410									17-Dec-08, One Path Pro
A1000	Start Milestone	0d	01-Jun-08*						◆ Start Milestone							
A1010	Design	50d	01-Jun-08	20-Jul-08	40	50	75			Design						
A1020	Build	100d	21-Jul-08	28-Oct-08	80	100	140				Build					
A1030	Test	35d	29-Oct-08	02-Dec-08	25	35	70						Test		
A1032	Root Cause Analysis	0d	02-Dec-08	02-Dec-08	5	10	20						I Root Cause Analysis	
A1033	Determine the Recovery Plan	0d	02-Dec-08	02-Dec-08	10	12	25						I Determine the Recovery Plan	
A1034	Execute the Recovery Plan	0d	02-Dec-08	02-Dec-08	10	15	25						I Execute the Recovery Plan	
A1035	Retest the Item	0d	02-Dec-08	02-Dec-08	15	20	35						I Retest the Item	
A1040	Ship	15d	03-Dec-08	17-Dec-08	10	15	20						Ship	
A1050	Finish Milestone	0d		17-Dec-08									◆ Finish Milestone	

Figure 7.4   Schedule with probabilistic branch activities added[2]

PROBABILITY THAT THE RISK EVENT OCCURS

The SMEs have assessed the probability that the article, which is the objective of this project, fails the test 30 percent of the time. That is, if we were to do the project 1000 times over and over again with the same risks, the test would fail in 300 of those projects. Assuming this value is close to correct requires our belief that the SMEs are not biased, an opinion that we need to validate through multiple interviews.

The simulation has to incorporate the possibility of the impact of failing the test 30 percent of the time—that is, in 30 percent of the iterations. This capability, which is included in most schedule simulation software programs, is called Probabilistic Branching. The logic of the schedule we have now created, which includes activities that are only 30 percent likely, is shown in Figure 7.5.

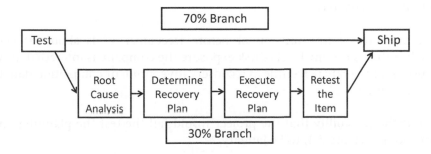

Figure 7.5   Pure logic diagram of the probabilistic branch

---

2   This chart uses Primavera P-5 scheduling software.

The Monte Carlo simulation program runs the schedule many times. Before each iteration the durations are selected at random for each activity including each recovery activity that has an uncertain duration. As before, if the activity exists in any iteration, the selected durations are used as if they were certain. For a probabilistic branch it is this existence that needs to be determined for each iteration, in this case for only 30 percent of the iterations chosen at random.

It is this statement, 'if the activity exists,' which is special with probabilistic branches. With the probabilistic branch shown above, the simulation program calculates the schedule up to and including the Test activity. Imagine that the program then flips a 70 percent–30 percent coin to determine if the four activities (the '30 percent probable branch' in our example) that are on the failure branch are in or out of the project for that iteration:

- If the coin turns up 'heads' (30 percent of the iterations) the recovery activities are in the project. Clearly they will add duration to the schedule as the organization investigates what might have happened, plans and implements the recovery plan, and retests.
- If they are not in the project (70 percent of the iterations), the durations of all activities along the recovery path are set to zero and they have no influence on the outcome of the project.[3]

The results for this project are shown in Figure 7.6.

INTERPRETING THE RESULTS—A BI-MODAL DISTRIBUTION

Notice that the probability distribution no longer has a simple single-mode shape. That is because of the probabilistic branch. Imagine that this histogram shows two somewhat-overlapping single-mode distributions:

1. The taller distribution on the left includes 70 percent of the iterations representing the same project performed over and over again, 3500 out of 5000 in this case, where the failure branch was not chosen and the failure activities' durations were set to zero.
2. The shorter distribution on the right includes 30 percent of the iterations or projects, 1500 in this case, where the failure branch was chosen and the failure activities' durations, chosen at random from their triangular probability distributions, were different from zero days.

The probability of finishing on or before December 17 is about 10 percent in Figure 7.6 but that is not the level of risk exposure the company is interested in. We have been focusing on the 80th percentile as a conservative, prudent completion date to adopt as a planning target.

- Without the possibility that the project would fail the test the planning target was estimated (see Figure 7.3) to be January 22.

---

3   In this case the particular software, Pertmaster® from Primavera assigns 100 percent likelihood to the Ship activity since it has more than one predecessor. In other software, for example, Risk+ from Deltek, the second successor of Test can have 70 percent likelihood. We can get the same effect with most schedule simulation software products.

- With the probability of failing the test set at 30 percent and the four new activities that recovery would entail, the 80th percentile has extended to March 30.
- Put another way, without the probabilistic branch the contingency needed for the organization's desired degree of certainty was 36 project days. That has extended to 103 days with the probabilistic branch.
- A final way to look at it is that, on top of a schedule that is 200 calendar days long, the contingency to P-80 of 36 days without the branch is an additional 18 percent. Recognizing the possibility of failing the test (the probabilistic branch) the contingency to the P-80 stretches to 51 percent. This change reflects the relationship between the target contingency of P-80 and the probability of success (70 percent). If a target below P-70 were used the differences would not be as dramatic.

Notice in Figure 7.6 that the cumulative distribution or S-Curve has a 'shoulder' at the 70 percent level in the picture. This reflects the SMEs' assessment that the product will pass the test 70 percent of the time and fail only 30 percent of the time and, of course, would change its position if the assessment of probability were otherwise. The position of the shoulder provides some insight into the importance of the degree of certainty the organization needs and whether this analysis affects that organization. The important differences in dates occur above the 70th percentile.

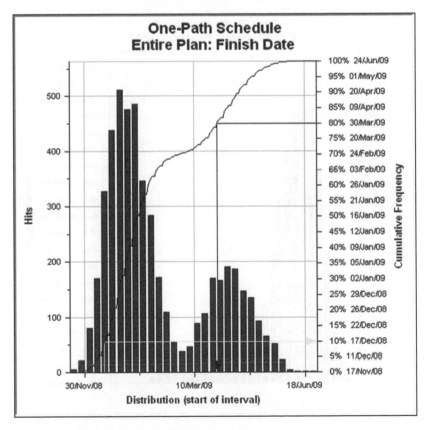

Figure 7.6   Simulation results for a probabilistic branch with 30 percent probability of occurring

Risk Mitigation that May Seem Counterintuitive

An organization that demands an 80th percentile schedule finds itself with a possible way out of the dilemma through a risk mitigation that may be somewhat counter-intuitive. If the organization can *take a little longer in the building of the product*, making a higher-quality product and introducing more testing of various units before the entire system is integrated and then system-tested, the probability of failure might be reduced from 30 percent to, say, 10 percent. Even if that failure is severe, perhaps with the same consequences as before (see Table 7.2), *a shorter schedule and earlier completion date may result at the 80th percentile*. This counterintuitive result occurs because the probability of failing the test is reduced to 10 percent if the shoulder occurs only at the 90th percentile level, above the target level of the 80th percentile.

In Figure 7.7 the duration of the Build activity has been increased by 20 days as have its three estimates (optimistic, most likely, pessimistic) and the probability of the branch has been reduced to 10 percent. (The analyst should confirm with the project team whether this is possible, and what parameters should be used.) The resulting distribution is shown below and is compared to the original one before risk mitigation.

Comparing the two scenarios on the same overlay chart, Figure 7.8, indicates what is happening.

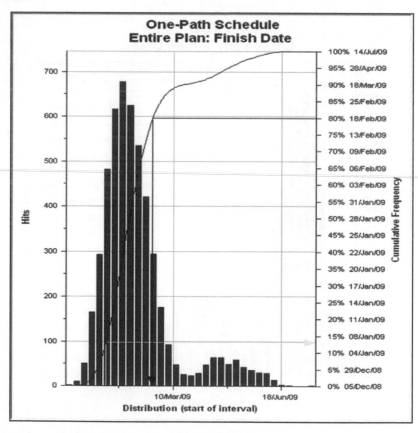

Figure 7.7   Risk mitigation build takes 20 days longer but reduces the probability of test failure to 10 percent

Adding 20 days to the Build activity and thereby reducing the probability of Test Failure to 10 percent actually *improves the schedule for an 80 percent organization* from March 30 to February 18 or about 40 days. Taking more time to produce a product that will more-likely qualify at testing actually saves time for a conservative company.

Although the possibility of a discontinuous event such as failing a qualifying test is not often included in a project schedule, it is something that has to be included in the risk analysis.

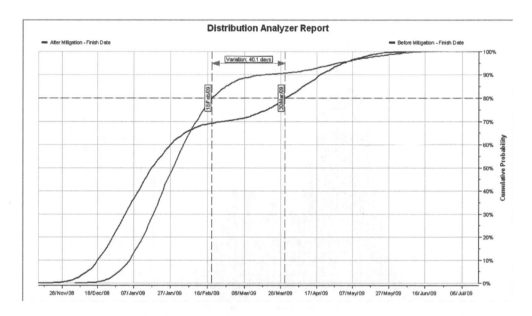

Figure 7.8   Effect of taking 20 more days in build to reduce the probability of test failure to 10 percent

Suppose there are Several Discontinuous Activities

Often in the later stages of the project there will be a series of tests that must be passed before delivery or a series of activities commissioning the plant before turnover. Each of these tests or commissioning activities may be failed with some probability, or else they would not be required. Interviews with the project participants and other SMEs develops data on the probability that these discontinuous events may occur and the impact on the project schedule if they do occur. What does this do to the shoulder of the cumulative distribution or S-Curve?

Let us specify three different tests. For simplicity, each of these has a single activity of 35 days rather than a four-activity recovery action as indicated above.[4] Each of these tests may fail, and we will introduce them in a cumulative way. As each is introduced the probability of failure will be the familiar 30 percent.

---

4   These recovery plans are not strictly probabilistic 'branches' but rather probabilistic activities, and they could be handled by the 'existence' and 'risk register' functions included in some simulation software.

With one branch the results exhibit the typical bi-modal distribution with the shoulder at 70 percent, the probability that the test will be successful according to the interviewees shown in Figure 7.9.

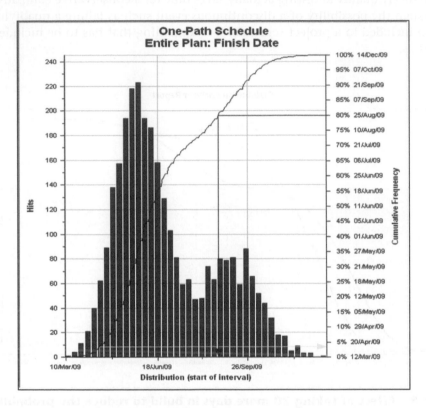

Figure 7.9   One probabilistic branch

With two probabilistic branches the results are as shown in Figure 7.10. The shoulder has now migrated down to about 50 percent so even a risk-neutral company that espouses the P-50 as their goal will be in the failure group of probabilities.

Finally, with three branches the probability of being in the failure area has increased to about 65 percent and the shoulder is rather indistinct, but may be at 35 percent as shown in Figure 7.11.

When the project encounters one, two and then three 30 percent branches the likelihood that the project will avoid all of the failures becomes small and the possibility of delay becomes greater. The overlay chart (Figure 7.12) shows the four scenarios together starting with no probabilistic branch on the left-most S-curve. From there branches are added one at a time and the resulting curves migrate to the right indicating that the overall schedule risk at all probabilities increases. The shoulder declines from 70 percent to about 35 percent and the distribution tips over to the right so an 80th percentile company's completion date slips, in this case, from June 19 to November 11.

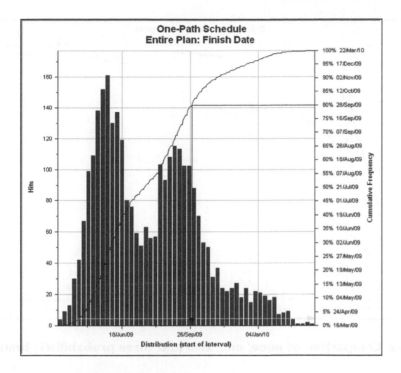

Figure 7.10 Two probabilistic branches

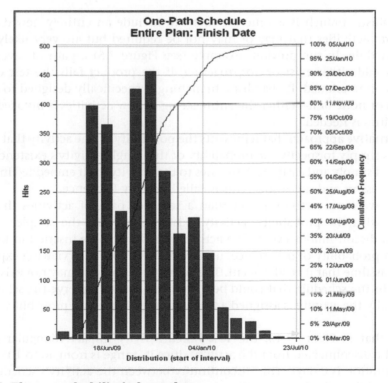

Figure 7.11 Three probabilistic branches

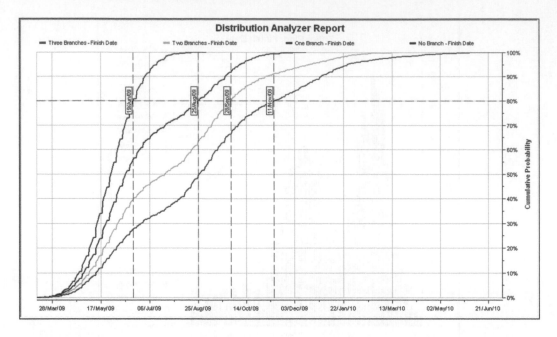

Figure 7.12 Comparison of none, one, two and three probabilistic branches

Activity Existence May Represent Probabilistic Events

The probabilistic branch is a structure that can include an entirely new direction and set of several activities that may or may not be required but are very likely not in the original schedule. In the previous example (see Figure 7.5) a path of several recovery activities would either exist or not, together, if the product fails the test or the plant cannot be commissioned. Probabilistic branching is specifically designed to activate the entire path or none of it—either the entire string of new activities is invoked or none of those activities is needed.

An alternative approach that represents the possibility of one activity that may or may not occur is simply to specify the probability of that single activity's existence. Existence is introduced simply and easily as it attaches to the activity itself embedded in a path with other predecessors and successors. This is called Activity Existence.

If there were one activity rather than a series or path of activities that would be introduced with the probability, activity existence may be the simpler approach to specifying a discontinuous event. The activity may have to be inserted in the schedule. It is given a probability of occurrence, usually derived from interviewing experts. In this case, let us assume that it is 30 percent. The activity that may or may not exist is assigned a duration in the schedule that could be 0 days in order to preserve the schedule results *ex ante*. Finally the activity is assigned a three-point estimate and probability distribution if it occurs.

Notice that in Figure 7.13, the Test Failed activity impact triangular distribution appears to be disconnected from the activity since its range is from 40 to 105 days but its schedule duration is 0 days. This discontinuity between the activity's schedule duration and its probability distribution will cause some simulation programs to warn the user that the duration is outside of the risk range and that the most likely duration is not the same as the schedule duration. In this case we understand those facts and actually caused them to happen, so the warnings can be examined but ignored.

In simulation, the activity 'fires' at random, based on the probability that is assigned. This is just like the probabilistic branch in which the activities all fire or not together. So, for our example above, in three out of ten iterations the test failure would occur and in the other seven iterations it would be given a value of zero. In those seven iterations when the activity does not fire (occur) the predecessor activity, Test, just leads directly to the successor activity, Ship, as if the probabilistic activity were a simple milestone taking no time. The results are shown in Figure 7.14.

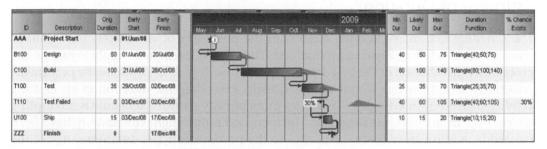

Figure 7.13 Simple existence representation of discontinuous event

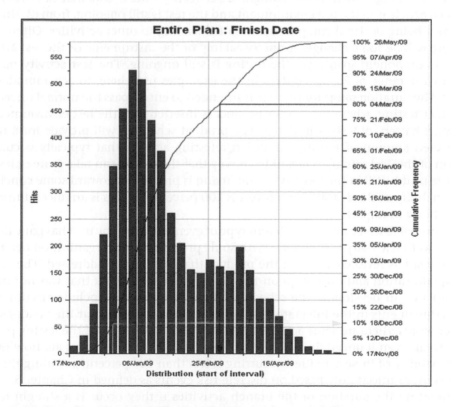

Figure 7.14 Results of probabilistic existence of a test failure

Clearly one could use the activity existence option instead of the probabilistic branch if the activities that occur in the discontinuous event can be summarized into one activity. Hence, the path with four activities that occur only if the article fails the test could be summarized into a Test Failed activity with some range of duration. This approach may be adequate if there is not much information about what would happen upon failure of the test, or if there are so many discontinuous risk events that it is cumbersome to model them all with branches.

When to Use a Branch Versus Extending the Three-point Estimate of Duration

The schedule risk analyst is often faced with the possibility of an unlikely but dramatic scenario such as failing a crucial product test. The analyst is tempted simply to extend the probability distribution of the base activity, in this case Test, to incorporate the pessimistic possibility of a delay due to test failure. Clearly, extending the duration of an existing activity instead of inserting new activities in an already crowded schedule is a tempting alternative. In the case just suggested, the increased range might conceivably have been created by adding an additional 105 days to the pessimistic upper duration range of the activity's duration.

Modeling the failure branch correctly is important and preferable to extending the Test activity's range. We have to distinguish between (1) the events that can occur while the tested article is in the test environment and the test is still ongoing, from (2) the event of the test being declared completed and failed leading to other activities. Often during test there will be some changes, from 'tweaking' of the instruments or the test article to swaps of component spares, yet the testing is still ongoing. The Test activity has been given a wide range of 25–35–70 days in the examples used here, to accommodate this activity. The duration range for the Test is intended to encompass the normal uncertainty of product testing, where the unit or product is inserted into the test environment and a complex battery of tests is initiated. The prudent scheduler will provide for a normal or expected amount of testing/calibration/retesting activity that typically occurs. The duration of the Test activity is meant to convey that the testing can take more or less time while the unit is still in the test environment and is proceeding toward some conclusion. The probability of the distribution on Test is 100 percent and this is an uncertainty based on variability in impact.

The probabilistic branch is a different type of event entirely. The unit has gone through testing and that activity is complete. With some probability, the organization determines that the test has revealed a flaw in the product that needs to be addressed. The unit has failed qualification and cannot go on to Ship without some work that was not initially contemplated in developing the schedule. These activities can include anything from minor adjustments in the integrator's shop to major re-work back at the vendor's shop. In other words, there are some activities that would not be needed if the testing proceeds in a normal, even if uncertain, way toward success, but these activities are now needed. The probability of these activities' occurring is less than 100 percent, so failing the test is a risk event, or uncertainty based on discrete risk events as defined in Chapter 5.

Estimating the duration of the branch activities if they occur is a straight-forward estimating job, is no harder than developing the schedule in the first place. What is new is estimating the probability that the branch will occur at all. Interviewees often will downplay the possibility of failure for one or several reasons:

- failure has not been included in the base schedule so it has been ignored. The anchor is effectively that probability is 0 percent;
- failing the test is such bad news that interviewees cannot give it recognition;
- corporate policy is for 'success-oriented' schedules and predicting failure, even probabilistically, is not allowed by the culture.

Alternatively, some interviewees may over-estimate the probability of failure if they have personally experienced such a failure within recent memory (and memories of failures tend to hang on for a period of time). The risk analyst must be alert to any type of bias in representing the probability of these risk events.

Summary to Probabilistic Branching

This chapter introduces the phenomenon of risk events that may or may not occur, but if they occur their impact on the schedule is uncertain. We used an article's failing a qualifying test to represent an event that may occur with some probability. Test failure may be unexpected and is usually not included in the schedule, however, so we have to add some activities. We gave those activities a duration of 0 days in the schedule to keep them from affecting the schedule's completion date. Then we supposed that the experts estimate that there is a 30 percent chance that the article will fail the test and that some recovery activities that are otherwise unneeded will be required.

The recovery activities formed a path that is called a probabilistic branch. This branch contains a short path of several activities that is either needed or not. It will be triggered or fired randomly during a simulation in 30 percent of the iterations if that is the probability specified by the SMEs during the interviews. The resulting histogram is often bi-modal with the iterations representing a successful test making up one mode and the iterations during which the unit failed the test making up a second mode. The cumulative distribution or S-curve exhibits a shoulder at the percentile of successful tests, 70 percent in our example.

The shoulder at 70 percent provided an organization that requires an 80 percent-likely schedule with a somewhat counter-intuitive risk mitigation opportunity. Taking more time in the front end with more exhaustive unit testing, for instance, may result in a shorter project. If the mitigation is believed to lead to a reduction in the assessed probability of failure to a low number, in our example 10 percent, the P-80 date may be earlier than before mitigation.

Often there is a series of tests at the end of projects and each one could be successful or not. Hence there may be the need for multiple probabilistic branches to analyze the risk to the schedule. With several branches, the shoulder level declines (in our example from 70 percent to about 35 percent), and even the 50-50 company is caught in the failure mode for at least one test. With multiple tests and branch possibilities, one is cautioned not to over-model by building branches with insignificant recovery processes or branches within branches representing multiple-failing of a single test.

If a single probabilistic activity may or may not exist, or if a branch can be summarized by a single activity, a variation on the probabilistic branch is the activity existence approach. While the branch groups several activities together and they all occur or they all do not, activity existence can handle a single probabilistic activity. This approach may be useful and simpler to handle if there are many uncertain risk events leading to discontinuous impacts and the results of each can be specified by a single activity.

Finally we distinguished between discontinuous events such as when the test is declared to be complete and the article failed the test from the normal uncertainty that occurs during testing and makes testing an uncertain duration activity. This distinction is between uncertainty and risk events as described in Chapter 5. The discontinuous events require either probabilistic branches or activity existence treatment that can exist with a probability less than 100 percent. Uncertainty about how long the article may remain in test while still emerging successful is handled by broadening the probability distribution of the test activity. In interviewing experts we need to be able to distinguish between these two alternatives since they are handled differently in modeling and simulation.

References

PMI (2004). *A Guide to the Project Management Body of Knowledge*. Newtown Square, PA, Project Management Institute.

8 *Using Risks to Drive the Analysis and Prioritize Risks: Introducing the 'Risk Driver' Method*

Introduction—Some Limitations of the Traditional Schedule Risk Analysis

This chapter introduces a new approach to analyzing schedule risk and contrasts it with the traditional approach that is presented in the earlier chapters of this book.

The traditional practice of schedule risk analysis represents schedule risk by placing uncertainty on activity durations using three-point estimates of low, most likely and high days' durations and selecting appropriate probability distributions before conducting a Monte Carlo simulation.

Some would say that a weakness of the traditional method is that it focuses on impacts rather than risks. This is partly true because the ranges of possible activity durations summarize all of the various risks that influence the activity's duration. In some sense the identity of the actual risk causing the uncertain duration is masked by combining all risks affecting one activity into one three-point estimate.

There are some limitations to the traditional approach that result from using the three-point estimate to place risk directly on the activity durations:

- The linkage between the qualitative risk analysis which prioritizes the risks and the quantitative risk analysis which applies risk to activity durations is often vague and imprecise. The risk register, including a prioritization of risks that affect the schedule, is consulted before developing the three-point estimate in qualitative risk analysis, but we lose the ability to identify which specific risks drive the activity durations' uncertainty.
- Usually risks are characterized by their probability of occurring as well as the range of duration impacts that are possible if it were to occur. In the traditional method of schedule risk analysis, however, the only uncertainty is the degree of impact on the activities' durations represented by the three-point estimates of the activity durations. There is no place to insert the probability of the risk's occurring since these duration ranges are presumed to occur with a 100 percent certainty. In this way the traditional approach can represent the uncertainty and ambiguity types of risk (see Chapter 5). A three-point estimate that occurs with 100 percent certainty cannot represent risk events that also have a probability of occurrence that is less than 100 percent. The

concept of probability of a risk's occurrence, which is central to the concept of risk events, is used only in probabilistic branching or existence-type risks in the traditional approach.

- Management wants to identify the most important risks jeopardizing the schedule in order to focus their risk mitigation efforts efficiently. The traditional approach cannot sort out which risks might cause potential schedule overruns. That is because measures such as risk criticality and schedule sensitivity introduced in Chapter 6 identify only those activities and paths that stand out as being important to the determination of the completion date. Criticality and sensitivity will not indicate which specific risks are crucial. This is particularly confused in the traditional method of schedule risk analysis since some risks may affect several activities but risk criticality and sensitivity focus on individual activities and paths. The analysis of a risk's importance in the traditional approach described in earlier chapters can only be based on its impact on specific tasks and not on all tasks it may affect.

The New Risk Driver Approach

An alternative approach, called the risk driver approach, starts with the risks prioritized in the risk register and drives the activities' uncertainty directly from those risks. It addresses the criticisms bulleted above. In the risk driver approach the risks are identified and applied to all activities they affect. The risk interviews focus on risks, not on the impacted activity durations. The probability distributions for each activity are developed based on the probability and impact of all risks that are assigned as those risks may occur or not and, if they do occur, how they impact the activity durations. In this approach we model how each individual risk may affect the durations of one, several or many activities.

For the risk drivers approach:

- The uncertainty is associated with each risk, not with the activity that is affected by risks. This uncertainty includes both probability of occurring and impact on durations if it occurs, so it fits the concept of project risk well. These measures are explored and calibrated in percentage terms during the risk interviews.
- The risks are applied, or assigned, directly to activities. A risk can be applied to several activities and an activity can be affected by several risks. Thus we are modeling the way activity durations are affected by risks in real projects.
- Measures of importance such as sensitivity are applied to the risks themselves, not to the impacted activities. As such, these measures take into account all of the activities the risks affect as well as how those activities act through the schedule to influence the final schedule milestone. Hence, sensitivity analysis (tornado charts) list, in order of importance, the risks rather than activities. Normal measures of importance of activities such as criticality and sensitivity based on activities are still available.
- The total contingency reserve can be explained in terms of the risks that cause it. The most important risks can be identified for further analysis or mitigation. Mitigation actions focus on risks, not on activities, and this approach focuses on priority risks so that mitigation actions can be developed. For instance, a mitigation action might reduce the probability that a risk will occur. This reduction in probability can be put into the model and a new simulation conducted. The difference between the all-in

simulation and the risk mitigation scenario, say at the 80th percentile, can estimate the impact of the risk mitigation.

• The risk drivers method can be combined with other traditional risk analysis approaches. These include:

– Applying an 'estimating accuracy' probability distribution on activities to represent the quality of the information underlying the estimates of duration. For instance, each activity could be given a risk range of -5 percent and +10 percent representing a late-stage mature estimate of schedule durations.

– Probabilistic branches representing the possibility that a discrete risk could lead to a path of activities not usually included in the project. A common use of these branches is to represent the work needed if a test is a failure.[1]

The risk driver method is in some sense an extension of the probabilistic branching and activity existence methods introduced in Chapter 7, since it deals with individual risks as the driving forces affecting the overall schedule. The structure of the analysis is quite different from probabilistic branching, however. The probabilistic branch can represent a chain or path of activities that may occur with some probability but it occurs in one place in the schedule. The risk driver can affect the duration of many activities that are already present in the schedule, but with some probability that may be less than 100 percent.

Source of the Project Risks used in the Risk Driver Method

Most projects have a risk register compiled before quantitative schedule risk analysis is conducted. The risk register includes identified risks that have been assessed and prioritized for their importance to the project. Historically risk registers contained only threats to the schedule—that is, if the risk were to occur it would delay the project in the opinion of the Subject Matter Experts (SMEs) who compiled the list. These days risk registers also include opportunities—that is, if they were to occur the project schedule would be shorter than planned. The risk drivers approach allows both opportunities and threats.

Sometimes the risk register separately identifies risks that are thought to have an impact on the project's schedule, cost, scope and quality. It is the risks thought to be important influences on the project's schedule that we are concerned about in this book. The risk register usually lists the probability that each individual risk will occur and, if it does occur, the impact on the project.[2] This information will form the basis of the risk drivers approach to project schedule risk analysis, although we typically conduct quantitative risk interviews to gather specific information needed for this quantitative approach.

Since the risk register already exists it is natural to use it as the basis of the schedule risk analysis. Doing so makes explicit the connection of the qualitative risk analysis process with the quantitative risk analysis process that helps compute the overall project

---

1   See Chapter 7, which introduces probabilistic branches.

2   Notice that the impact is assessed without explicit reference to the schedule structure or to the number of activities the risk will affect—the qualitative risk register assessment is not as rigorous as the risk driver quantitative assessment of the impact.

schedule risk, the contingency required and its own prioritization of risks within the structure of the project schedule.

The Risk Driver Method Preview

In the risk driver method:

- The connection to and reliance on a good risk register is made explicit. The risks are specified explicitly within the analysis based on the same concepts that make up the prioritization of risks within the risk register. This connection between qualitative and quantitative risk analysis has not always been as strong as it should have been.
- Risks are characterized by their probability of occurring and impact on schedule activities if they do occur. Their impact may be positive (opportunity) and/or negative (threat) and both may be combined in the distribution. This is consistent with the definition of a project risk: an uncertain event or condition that, if it occurs, has an impact on at least one project objective.[3]
- Risks are assigned specifically to each activity they influence. One risk can affect several activities, in which case the influence is perfectly correlated in the schedule. This is seen because: (1) if the risk occurs in one iteration of the simulation it occurs for all activities to which it is assigned, and (2) if its impact is a strong threat (opportunity) for one it is a strong threat (opportunity) for all affected activities. Thus the durations of each activity to which the risk is assigned 'move together,' the description of correlation.[4]
- An activity can be influenced by several risks. If several risks that are assigned to one activity occur during an iteration the duration is influence by all assigned risks.[5]
- At the end of the analysis, the overall project risk is measured in terms of a contingency reserve of time that would be required from the scheduled completion date, just as in traditional schedule risk analysis. However, using the risk drivers approach, that contingency, say 250 days, can be decomposed to show the marginal contribution of each risk. This occurs as we take out each risk starting with the most important risks and finally eliminating all or most of the risks and getting back to the schedule date. As each risk is taken out (by giving it a probability of 0.0 percent) we remove its influence from all of the activities to which it is assigned. Comparing the schedule risk results both before and after a risk is taken out measures the risk's marginal impact on the contingency reserve of time. This process provides the project manager and the organization with the impact in days of specific risks and prioritizes the risks that need to be addressed in risk mitigation.
- As risk mitigation is being planned, the probability and/or impact of the risk, or its assignments to various activities, can be altered and the analysis re-run. The comparison of the results, say at the 80th percentile, before and after the risk mitigation

---

3   Guide to the Project Management Body of Knowledge (PMBOK® Guide), 3rd edition, 2004, Project Management Institute, p. 238.

4   Chapter 10 describes correlation between activity durations.

5   If the risk does not occur for an iteration it is given a value of 1.0, so it has no effect on the multiplication of factors.

is modeled, indicates the potential benefit of the mitigation action. The cost of the mitigation can be compared to the time-impact of improving the project schedule to calculate a benefit-cost analysis of risk mitigation. This will help make intelligent choices within a complex project of which risks to mitigate first.

- The risk driver method is based on the notion that the risk drivers are very basic or root causes and are independent of one another. Thus the drivers themselves are not correlated, though the activities themselves may be correlated as mentioned above.
- The interviews are focused on the risks and stay at a summary level. This level is a good one because the interviewees think of risk at that level and they usually have more trouble specifying risks to individual activities in the traditional schedule risk approach.

The Risk Driver Method—Mechanics[6]

The risk driver method starts with a risk register. We will explain and demonstrate the method using:

- individual risks applied to an activity. These risks show how the activity duration reacts to risks with different probability and impact assumptions;
- activities with multiple risks applied;
- a simplified schedule for the development of a new refinery.

Risk drivers are expressed in the traditional way that we describe risk events, with probability of occurring and impact if they do occur.[7]

- The probability that the risk driver will occur on this project is assessed during the interviews. The risks considered were deemed to be important during the qualitative risk analysis process that produced the risk register, so we expect that each risk has a significant probability of occurring. The interviews check and calibrate this expectation.
- Impact on activity duration if the risk does occur. This impact is expressed as a percentage of the activities schedule duration and is used as a multiplicative factor when applied to an activity's duration.
- We also collect information on the identity of the activities the risks impact, if they occur. This information is needed for the assignment of risks to activities in the schedule.

The multiplicative factor works like this: suppose a risk, say construction supervision may be scarce, is assigned to an activity like Site Work with a duration of 100 days. Perhaps the risk factor's range, determined during interviews, is .90, 1.00, 1.15. The risk's value might be selected at random to be 1.05 for a specific iteration. Applied to the 100-day duration, Site Work would have a 105-day duration for that iteration. If it does not occur for another iteration, the factor is 1.0 and the duration of Site work is 100 days.

---

6   In this chapter we are using the Risk Factors module from Primavera Pertmaster.

7   A major advantage of the risk driver method is that it can also represent an ambiguity or an uncertainty if the probability were set to 100 percent.

Specify the Risk—Risks that are 100 Percent Likely to Occur

To show how the risk drivers affect the activities' durations, we consider several general cases. The risk driver percentage is applied to a risk with a duration of 100 days to make it simpler to see the impact of risk drivers.

Consider the following risk (Figure 8.1):

	Description	Optimisitic	Most Likely	Pessimisitic	Likelihood
1.	Construction Labor Productivity May Vary	90%	100%	115%	100%

Figure 8.1   Risk: Probability 100 percent, impact range including opportunities and threats

This risk is actually an uncertainty—the existence of some value of labor productivity is not uncertain but its level is uncertain. The impact can take on different values when applied to activity durations. It is a strength of the risk driver method that it can encompass risks with a 100 percent probability, called uncertainties or ambiguities, but uncertain impact and that the impact can be both opportunities (factor < 100 percent) and threat (factor > 100 percent).[8]

Let us assign that risk to a particular activity with duration of 100 days. In a Monte Carlo simulation, the Construction activity to which risk #1, Construction Labor Productivity May Vary, has the following probability distribution (Figure 8.2).

Notice that the range is from 90 days to 115 days and that the most likely duration is 100 days, the duration in the schedule. However the probability that this activity's being 100 days or less is about 45 percent and the 80th percentile is 106 days. If this risk is the only one affecting Construction, and since its probability is 100 percent, Figure 8.2 looks like the traditional risk analysis three-point estimate on duration.

The probability distribution does not need to be centered on the duration in the schedule. Taking the next project risk (Figure 8.3), notice that interviewees believed that this, too, was 100 percent likely but that its impact is entirely a threat. Unlike the previous risk, here the impact starts at 100 percent in the optimistic case and increases from there for most likely and pessimistic.

The activity affected by this risk, Technology Design, has duration of 100 days. Figure 8.4 shows a distribution based on these parameters.

Specify the Risk—Risks that are Less Than 100 Percent Likely to Occur

Risks are described by their probability of occurring, and most risk events have a probability of less than 100 percent. If this is the case, the probability distributions for the activities affected exhibit a 'spike' where the risk factor is 1.0. The probability in the spike includes

---

8   In some risk register applications a risk is either a threat or an opportunity but not both. The risk shown here cannot be easily specified in the Risk Register application of Primavera Pertmaster, for example.

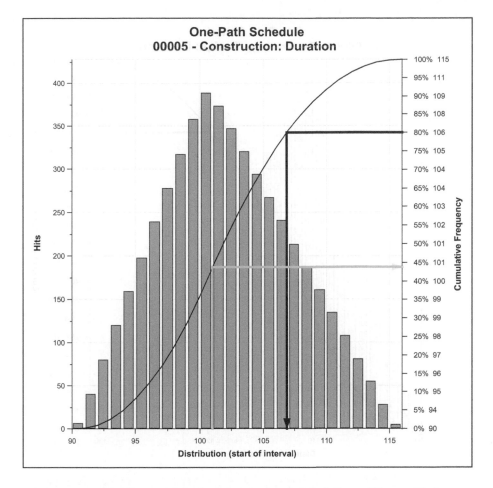

Figure 8.2   Construction activity with construction labor risk applied

	Description	Optimisitic	Most Likely	Pessimisitic	Likelihood
1.	Technology may be More Difficult than Planned	100.00%	110.00%	130.00%	100.00%

Figure 8.3   Risk: Probability 100 percent, impact only threat

the probability that the risk will not occur plus the probability that the value of the risk factor is 1.0 even if it were to occur. Figures 8.5, 8.6 and 8.7 show two such examples. Figure 8.5 shows the two risks that are illustrated in the subsequent figures. Notice that each of these risks has a probability of occurring, in the column labeled 'Likelihood,' that is less than 100 percent.

Figure 8.6 shows that Construction is affected by the construction productivity risk that is 30 percent likely to occur. In this figure the spike at 100 days includes mainly the 70 percent of the iterations on which the risk did not occur (notice the number of hits

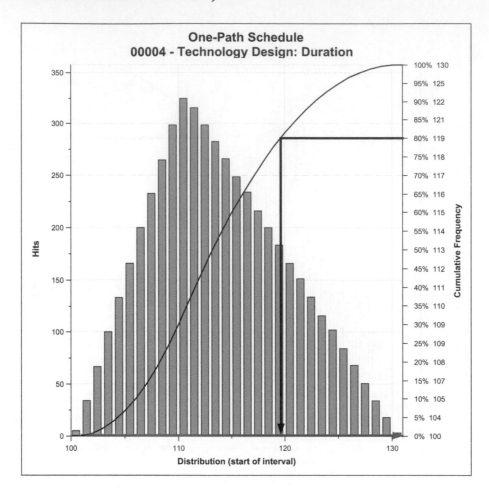

Figure 8.4 Design risk which is entirely a threat

	Description	Optimisitic	Most Likely	Pessimisitic	Likelihood
1.	Technology may be more Difficult than Planned	100.00%	110.00%	130.00%	60.00%
2.	Construction Labor Productivity May Vary	90.00%	100.00%	115.00%	30.00%

Figure 8.5 Two risks: Probability less than 100 percent

on the left-hand axis slightly exceeds 3500 which would be 70 percent of 5000 total iterations). This spike represents the iterations on which the risk did not occur and the risk factor is set at 1.0, plus some iterations where the risk occurred but the value of the risk factor was still 1.0.

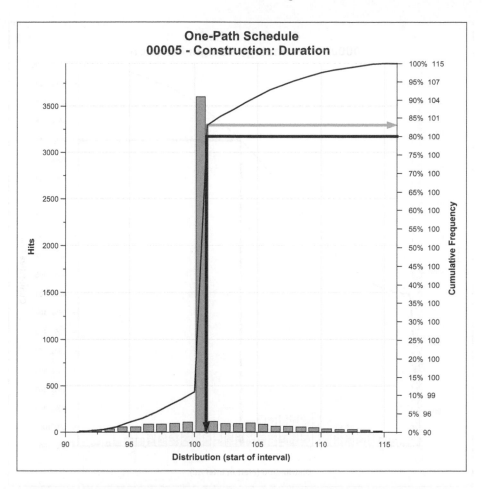

Figure 8.6   Activity with risk of 30 percent probability of occurring the spike contains slightly more than 70 percent of the probability

In Figure 8.7 Technology Design has a spike at its scheduled duration of 100 days that includes 40 percent of the probability (notice 2000 hits on the left-hand axis) because the risk occurs 60 percent of the time. The triangular distribution to the right includes 60 percent of the probability.

Activity With Two 100 Percent-likely Risks Assigned

An activity may have more than one risk applied to it. Let us consider two 100 percent likely risks that may be applied to a technology-related activity (Figure 8.8).

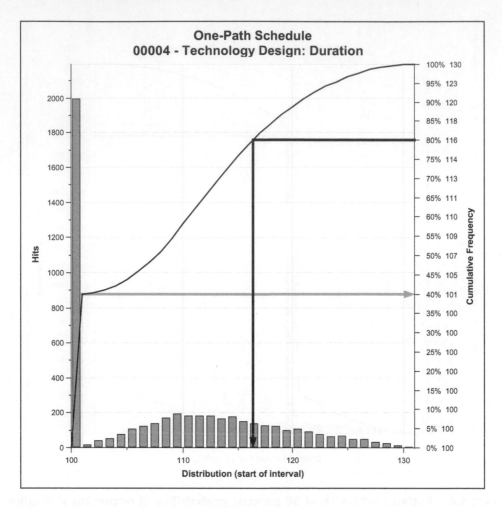

Figure 8.7 **Activity with a threat risk with 60 percent probability the spike contains 40 percent of the probability**

	Description	Optimisitic	Most Likely	Pessimisitic	Likelihood
1.	Technology may be more Difficult than Planned	100%	110%	130%	100%
2.	Technical Labor Productivity May Vary	90%	100%	115%	100%

Figure 8.8 **Two risks: Probability of 100 percent**

The results of assigning two risks to the same activity are shown in Figure 8.9. In this case, the mechanics are:

- The percentages for each risk are selected at random for an iteration. Say, for the difficult technology the value 1.2 (a threat) is selected and for technology labor the

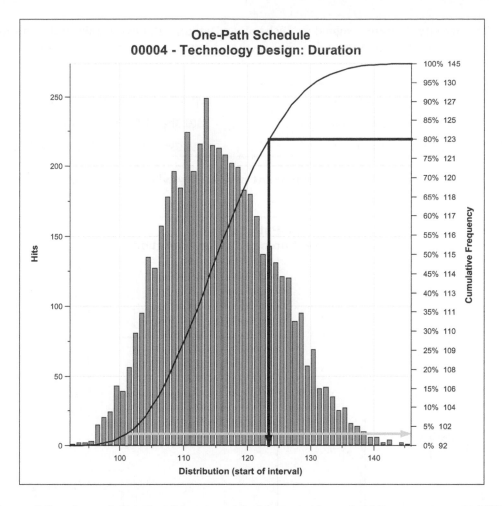

Figure 8.9   An activity that has two risk drivers assigned: 100 percent probability

value is .95 (an opportunity). Note that it is assumed in the risk driver approach that we are considering fundamental risks that are not correlated.

- The percentages are multiplied together. Thus, in this case, since each risk is 100 percent likely, the value would be 1.2 × .95 = 1.14. The activity duration would be multiplied by the factor 1.14 for that iteration.

Activity With Two < 100 Percent-Likely Risks Assigned

Notice that if one risk does not exist on a specific iteration its risk factor takes the value of 1.0. For that iteration its effect on the activity as the risk factors are multiplied is neutral. Hence, if only one risk driver occurs on a specific iteration, the value of the activity duration is completely determined by the assigned risks that do occur. Consider the following situation (Figure 8.10) in which we have reassessed the probability of the technology and technical labor productivity risk drivers to be less than 100 percent.

The activity they are assigned to, Technology Design, will have a spike at its duration (100 days) reflecting the probability that neither risk fires as shown in Figure 8.11.

	Description	Optimisitic	Most Likely	Pessimisitic	Likelihood
1.	Technology may be more Difficult than Planned	100.00%	110.00%	130.00%	40.00%
2.	Technical Labor Productivity May Vary	90.00%	100.00%	115.00%	50.00%

Figure 8.10 Two risks with <100 percent probability will affect the same activity

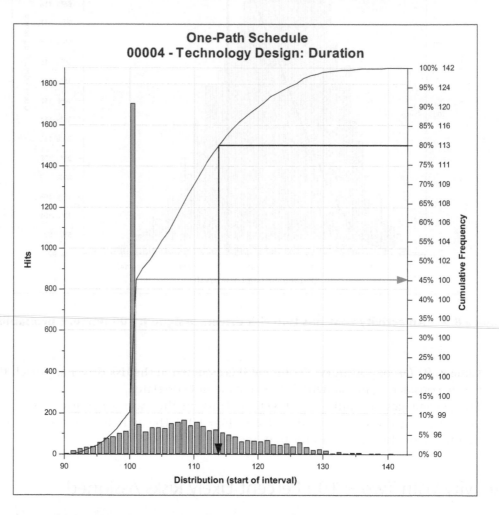

Figure 8.11 Activity with 2 risk drivers of probability of occurring of less than 100 percent

Sensitivity Analysis Using Risk Drivers

The sensitivity analysis can be conducted on either the activities (traditional method) or the risk drivers themselves (risk driver method). This is a major advantage of the risk driver method since the traditional method can only show sensitivity or criticality by activity, not by risk.

In this case the Tornado or sensitivity of the completion duration to the risk drivers is shown in Figure 8.12.

In this case, 'risk technology may be more difficult than planned', has a slightly lower probability than the 'risk technical labor productivity may vary' (see Figure 8.10). However, the difficulty of the technology risk has much more of an impact than the technical labor risk on the affected activity. This implies that more attention must be paid to the possibility that 'technology may be more difficult than planned' than to the 'technical labor productivity may vary' if an effective risk mitigation strategy is to be developed. Of course this presumes that an economically-feasible mitigation strategy can be found and implemented for the technology difficulty risk. One such strategy might be to examine other less challenging technical strategies. Another might be to seek out personnel who can handle better the challenging technology selected.

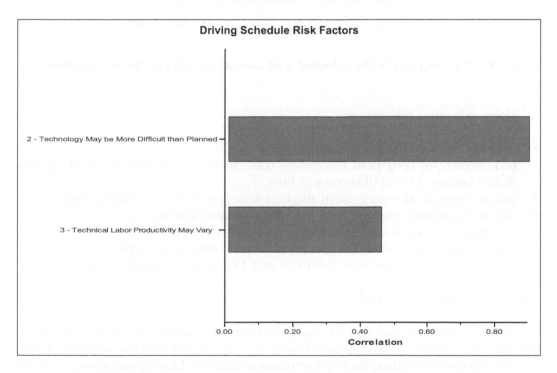

Figure 8.12 Tornado chart showing that the technology difficulty risk is more important than the technical labor productivity risk

Apply the Risk Driver Method to a Simple Refinery Construction Schedule

IDENTIFYING THE RISKS

In this simplified example (Figure 8.13) the risk register risks have been culled and consolidated to seven risks (Hulett and Whitehead 2007). For a real project there would be more risks, but a list of 30 or so important risks should cover the main concerns for the project.

1.	LLE suppliers may be busy
2.	Construction supervision may be scarce
3.	Suppliers may be busy
4.	Construction logistics may be different from plan
5.	Financing may be difficult
6.	Vendor reps may be scarce for commissioning
7.	Design productivity may differ from our expectations

Figure 8.13 Example of risks selected and summarized from the risk register

These risks cover the areas of:

- decision making—financing may be difficult;
- procurement of Long Lead Equipment (LLE) such as heat exchangers and heavy-walled vessels—LLE suppliers may be busy;
- procurement of other equipment and bulk materials—suppliers may be busy;
- construction labor—construction supervision may be scarce;
- construction—construction logistics may be different form plan;
- engineering—design productivity may differ from our expectations;
- commissioning—vendor representatives may be scarce for commissioning.

SPECIFYING THE SCHEDULE

A simplified, summary schedule of a refinery construction project is shown in Figure 8.14. Of course in a real project we will have several activities in each of these categories, but in the simplified schedule there are fewer activities so risks may be assigned to fewer, perhaps only one, activity.

APPLYING UNCERTAINTY—ESTIMATING ERROR

The first action we decide to take is to represent uncertainty in the accuracy of the activity duration estimates. This is not the final detailed schedule that will be developed once the contractor is on-board, and the information is viewed as being valid but not perfect. Suppose that during the interviews the accuracy of the estimates is determined to be within -10 percent and +20 percent across all activities. These ranges, applied to each of the activities' durations using Quick Risk, shown in the columns to the right of the Gantt chart in Figure 8.15, represent the uncertainty or estimating error. Applying a background risk range to each activity duration shows that the risk driver and more traditional methods can be combined if it is felt to be useful.[9]

Description	Rem Duration	Start	Finish	Total Finish Float	2009	2010	2011	2012	2013	2014
Project summary - us...	1510	01/Jun/09	19/Jul/13	-1						
Total Project	1510	01/Jun/09	19/Jul/13	-1						
Project Start	0	'01/Jun/09		-1	01/Jun/09					
Preliminary Authorization	30	01/Jun/09	30/Jun/09	0						
FEED Design	350	01/Jul/09	15/Jun/10	0						
Authorization	30	16/Jun/10	15/Jul/10	0						
Balance of Procurement	650	15/Aug/10	25/May/12	80						
Vendor Data Available	500	03/Nov/10	16/Mar/12	0						
Detailed Design	570	29/Oct/10	20/May/12	0						
Procurement of LLE	775	27/Apr/10	09/Jun/12	65						
Site Work	100	14/Sep/10	22/Dec/10	0						
Construction before LLE	600	23/Dec/10	13/Aug/12	0						
Construction after LLE	230	14/Aug/12	31/Mar/13	0						
Commissioning	110	01/Apr/13	19/Jul/13	0						
Project Completion	0		19/Jul/13	0						19/Jul/13

Figure 8.14 Simplified refinery construction schedule

ID	Description	Rem Duration	Start	Finish	2009	2010	2011	2012	2013	2014	Minimum Duration	Most Likely	Maximum Duration	E
SUMMAI	Project summary - us...	1510	01/Jun/09	19/Jul/13										
00001	Total Project	1510	01/Jun/09	19/Jul/13										
00002	Project Start	0	'01/Jun/09		01/Jun/09									
00003	Preliminary Authorization	30	01/Jun/09	30/Jun/09							29	30	33	
00004	FEED Design	350	01/Jul/09	15/Jun/10							333	350	385	
00005	Authorization	30	16/Jun/10	15/Jul/10							29	30	33	
000051	Balance of Procurement	650	15/Aug/10	25/May/12							585	650	780	
000052	Vendor Data Available	500	03/Nov/10	16/Mar/12							475	500	550	
000055	Detailed Design	570	29/Oct/10	20/May/12							542	570	627	
00006	Procurement of LLE	775	27/Apr/10	09/Jun/12							736	775	853	
000062	Site Work	100	14/Sep/10	22/Dec/10							95	100	110	
00007	Construction before LLE	600	23/Dec/10	13/Aug/12							570	600	660	
000072	Construction after LLE	230	14/Aug/12	31/Mar/13							219	230	253	
000075	Commissioning	110	01/Apr/13	19/Jul/13							105	110	121	
00008	Project Completion	0		19/Jul/13						19/Jul/13				

Figure 8.15 Refinery schedule with background risks applied

---

9   Other traditional methods such as probabilistic calendars, probabilistic branching and conditional branching can be used as well.

The completion date of the nominal schedule is July 19, 2013. The result of the simulation of this schedule with just the traditional background duration risk range (-10 percent, +20 percent) on every activity, is shown below (Figure 8.16). With just background risk the 80th percentile is September 9, 2013. Notice that it is about 5 percent likely that this project will finish on the nominal completion date of July 19, 2013 or earlier, even with just background risks applied.

ASSESSING THE PROBABILITY AND IMPACT FACTOR FOR THE RISKS

Now, let us suppose that we have interviewed specialists in the project's planning, execution and risk, and have collected the probability and impact information for each risk that is shown in Figure 8.17.

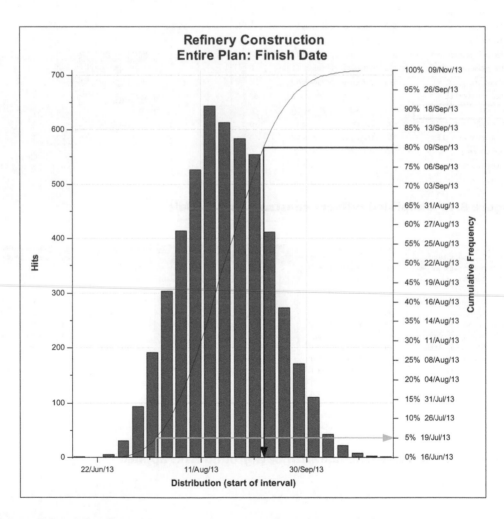

Figure 8.16 Schedule risk with only estimating risk applied

	Description	Likelihood	Dur Min	Dur Likely	Dur Max
1.	Financing may be difficult	20.00%	140.00%	200.00%	300.00%
2.	Design productivity may differ from our expectations	100.00%	90.00%	105.00%	120.00%
3.	Balance of procurement suppliers may be busy	50.00%	110.00%	115.00%	125.00%
4.	Construction Supervision may be scarce	50.00%	100.00%	110.00%	125.00%
5.	Vendor Reps may be scarce for Commissioning	35.00%	95.00%	115.00%	140.00%
6.	LLE suppliers may be busy	70.00%	105.00%	125.00%	140.00%
7.	Construction logistics may be different from plan	100.00%	90.00%	105.00%	115.00%

Figure 8.17 Risks probability and impact derived from the risk interviews

Importantly, the risks are described by both their probability of occurring and their impact if they do occur. Two of these risks, Design Productivity and Construction Logistics, are uncertainties with 100 percent probability of occurring and impact ranges, expressed in multiplicative factors. (The background estimating uncertainty, implemented in this example with traditional three-point estimates, could be implemented instead by a risk factor with 100 percent likelihood and .90, 1.00 and 1.20 factors that is applied to all activities.) The other risk factors are risk events with probabilities less than 100 percent and impact ranges.

In interviews some SMEs may not have opinions about each risk. In addition we may get different opinions on the parameters from several SMEs. The risk analyst is faced with the task of distilling the information down to the data inserted in this table, and there are two general problems:

1. The assessment of the likelihood of occurring is often placed too low by interviewees. Individual respondents may have difficulty expressing sufficient probability on the risks, particularly if the SMEs are inexperienced in risk analysis or the risk is particularly severe.
2. The range of impacts as reported is often too narrow. This may occur if the SMEs are inexperienced with risk analysis exercises on this or similar projects. Interviewees naturally wish to avoid the extreme values (in statistical terms, SMEs may find it difficult to report the 'outliers') that can be attached to a risk.

These common biases have to be recognized and can be addressed by the risk analyst before using the data to drive the simulation.[10]

ASSIGNING THE RISKS TO THE ACTIVITIES

Once the data are assembled, the risks have to be assigned to the activities. In the interviews the SMEs are asked which activities are affected by the risks. There are often risks that are assigned to more than one activity and activities with more than one risk assigned. In this simple example (Figure 8.18) we have made the following assignments:

---

10   See Chapter 5 for a more complete discussion of data collection for risk analysis.

Activities	Risks Assigned						
	LLE Suppliers	Construction Supervision	Suppliers Busy	Construction Logistics	Financing Difficulty	Vendor Reps Busy	Design Productivity
Preliminary Authorization					X		
FEED							X
Authorization							
Procurement of LLE	X						
Balance of Procurement			X				
Vendor Data Available			X				
Detailed Design							X
Site Work		X		X			
Construction before LLE		X		X			
Construction after LLE		X		X			
Commissioning						X	

Figure 8.18 Table of risk drivers and their assignments to schedule activities

Each risk is assigned to at least one activity and Site Work and Construction activities are affected by two risks, supervision and logistics. In real project risk analyses using the risk drivers approach some activities are affected by five or even more risks.

MONTE CARLO SIMULATION USING RISK DRIVERS AND BACKGROUND RISK

After the risks are assigned to the activities they might impact, the simulation is run with both the estimating error or ambiguity risks (if any) and the specific risk drivers. Some of these drivers may be uncertainties and some risk events. During simulation the steps for each iteration are:

1. The activity duration is selected at random from the background probability distribution (if assigned). In this case we assumed this distribution to be optimistically -10 percent and pessimistically +20 percent of the original duration with the original schedule duration assumed to be the most likely duration.
2. For each risk driver an impact factor is selected at random from the distribution in the table, shown in Figure 8.17.
3. For any specific iteration the risk driver is selected at random either to exist or not based on the likelihood derived from the interview, also shown in Figure 8.17. If the risk exists for that iteration its factor is the one selected in step 2. If it does not exist its factor is assigned the value of 1.0.

4. If the risk driver occurs for that iteration (see step 3) the activity duration for that iteration (see step 1) is multiplied by the risk factor (selected for that iteration in step 2) for the activities to which it is assigned.
5. If more than one driver assigned to an activity exists for the iteration, the selected impact factors (see step 2) are multiplied together and the resultant factor is multiplied by the duration of the activity selected from its background duration distribution (step 1). The result of step 5 is the duration of the activity for that iteration.
6. The project dates, including the overall project completion date, are calculated for each iteration using critical path method (CPM) rules as if the durations calculated as above are known with certainty. Of course, because the factors and duration are not known with certainty, the program computes the project in many iterations where the durations are similarly calculated.

The results of the simulation of the refinery construction project with estimating error and all of the risk drivers active and assigned as indicated in Table 8.1 are shown in Figure 8.19.

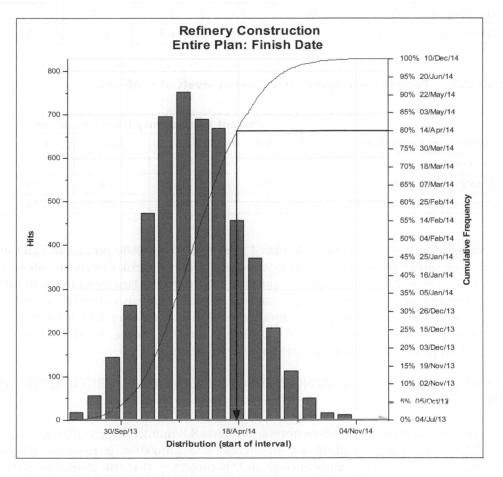

Figure 8.19 Schedule risk results with all risks considered

The results from these two simulations (estimating error alone and estimating error plus risks) are compared with the deterministic results and each other in Tables 8.1 and 8.2.

Table 8.1   Summary results from the schedule risk analysis

Summary Results—Total Refinery Project Schedule Risk Analysis					
Deterministic Date	P-10	P-50	P-80	P-90	
19-Jul-13					
Estimating Error Only	26-Jul-13	22-Aug-13	09-Sep-13	18-Sep-13	
Estimating Error plus All Risks	02-Nov-13	4-Feb-14	14-Apr-14	22-May-14	
	P-10		P-80	P-90	Range P-10 to P-90
	(Months from Deterministic)				(months)
Estimating Error Only	0.2	1.1	1.7	2.0	1.8
Estimating Error plus All Risks	3.5	6.6	8.8	10.1	6.6

Table 8.2   Percent contingency to different levels of confidence

	Percent Contingency from Beginning			
Duration from Project Start	P-10	P-50	P-80	P-90
49.6 months				
Estimating Error Only	0%	2%	3%	4%
Estimating Error plus All Risks	7%	13%	18%	20%

Notice that the background risk adds 1.7 months at the P-80 point and exhibits a 1.8-month spread between P-10 and P-90 points. When the risk drivers are added the risk at the P-80 is 8.8 months and the spread from P-10 to P-90 increases to 6.6 months. The percentage of contingency reserve implied by these results is indicated in Table 8.2, where the base is the original project duration of 49.6 months. If a schedule contingency reserve were to be added to the current completion date that reserve would be 8.8 months or about 18 percent of the total project baseline duration.

WHERE IS THE RISK? EXPLAINING THE CONTINGENCY AND PRIORITIZING THE RISKS

If the project manager or client does not want to add a 8.8-month (18 percent) contingency to achieve an 80 percent likelihood of success, risk mitigation is required. (Happily the analysis has been done early enough in this project so that risk mitigation may be productive.) The question is: which risks should be the focus of our attention?

Traditional schedule risk analysis, which starts with the three-point estimate on the activity durations, uses a risk criticality or sensitivity analysis measure to identify the activities that are most likely to be important in determining the variability in the project completion date (see Chapter 6). As a guide to risk mitigation these measures tell which activities are most in need of attention but not which risks are crucial to project schedule success. Measures of activities' importance to the result do not focus on the risks themselves. Hence:

- An activity may be risky because of several different risks and the most effective mitigation steps needed may not be clearly indicated. Mitigating one of the risks is likely to be only partially effective in reducing the uncertainty of the activity's duration. It may take mitigating several risks to reduce that uncertainty effectively.
- Highlighting the individual activities that are riskiest may understate the true importance of the underlying risks since those risks may impact several activities. Mitigating a risk that affects several activities' durations may have a greater benefit than it appears when priority is evaluated activity by activity.

With the risk drivers method of modeling the way the risks themselves affect activity duration uncertainty we identify the risks that drive the result rather than the activities that are involved. In this sense the activities are only the vehicle through which the risks act on the project schedule. Hence, the risk drivers method assesses the impact of the risks, acting through all of the activities to which they are assigned, on the total project outcome. In this way we can get a total impact of the risk on the schedule. This is done by:

- performing a Monte Carlo simulation with all of the risks included;
- subtracting the risks in priority order from the overall project risk analysis. This is done by setting the risk's probability to zero and conducting a simulation;
- determining the improvement to the schedule for the highest-priority risk by comparing the overall simulation result, say at the P-80 level desired by the project stakeholders, with the simulation's result at P-80 with the risk's probability set at zero for the highest-priority risk.
- continuing to subtract risks in priority order and compare each new simulation that omits the next-most important risk from the simulation before at the P-80. Taking out the next most important risk will save days at the P-80 and those days are arguably assignable to the risk taken out.

When all risks, including the background risk in this case, are removed one at a time we get the deterministic date and have 'explained' the total contingency at the P-80 level in terms of the risks themselves, rather than the activities that are impacted by the risks.

Management appreciates the ability to see the number of days each risk contributes to the overall contingency needed. This risks-focused (contrasted to the activity-focused) approach also leads management directly into discussions about risk mitigation, which is the most productive next step in the risk management process:

- If management does nothing, the analysis becomes a forecast of the project results using the current plan.

- If management turns to effective risk mitigation starting with the highest-priority risks first, the analysis will have served its purpose to cause the plan to change for the betterment of the project's schedule.
- The impact of an individual risk on the project's overall schedule risk will be determined by the risk's characteristics:

  - Its probability of occurring;
  - The range of impacts, multiplicative in percentage terms, that the risk will have on activity durations if it occurs;
  - The number and criticality of the risks to which it is assigned.

For the simple project we have used as an illustration of this technique, most of the few risks are assigned to only one activity. Because of this, the sensitivity of the project result to individual risks closely parallels the sensitivity of the result to individual activities.

In real projects, risks typically affect several activities' durations, so taking the risk out of the schedule neutralizes its effect on several activities simultaneously. The exercise of taking risks out of the analysis allows us to see the total impact of the risk on all activities it affects.

SELECTING THE RISKS TO ANALYZE AND PRIORITIZING THE RISKS

The risk factors tornado chart shown in Figure 8.20 indicates the sensitivity of the overall project duration to the individual risks. It shows that two risks that affect the Construction and Site Work activities are considered most important, followed by the LLE Suppliers' risk.

Figure 8.20, which focuses on the individual risks, is contrasted to the next chart, the typical sensitivity chart that focuses on the activities.

Contrast the information provided in these two tornado charts. Figure 8.20 states clearly that construction supervision is the problem, followed by construction logistics. These are actual risks. Figure 8.21 says that construction activities are risky, but management needs to get one layer deeper to the actual risks in order to find out why that is so.

These sensitivity charts indicate that a focus on the risk that the 'Construction Supervision may be Scarce' is the most important risk. However, it does not indicate how many days this risk contributes to the contingency reserve, a duration extending from July 19, 2013 to April 14, 2014.

HIGHLIGHTING AND CALIBRATING THE RISKS IN PRIORITY ORDER

We take out the first risk, 'Construction Supervision may be Scarce' by setting its probability to 0.0 percent. (Of course risk mitigation will never be this effective, so we are making an extreme assumption for the purpose of prioritization.) The results in Figure 8.22 show that, without that risk, the P-80 date is March 6, 2014. This is contrasted with the P-80 of April 14, 2014 with all risks considered. It indicates that this risk contributes over 1 month to the 8.8 months of contingency to the P-80 level of confidence.

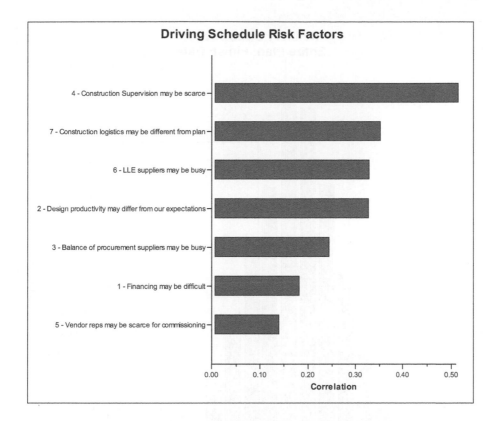

Figure 8.20 Risk driver sensitivity (tornado chart) showing priority risks

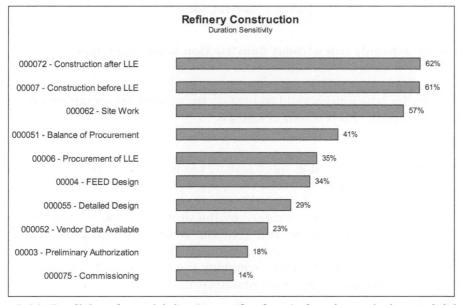

Figure 8.21 Traditional sensitivity (tornado chart) showing priority activities

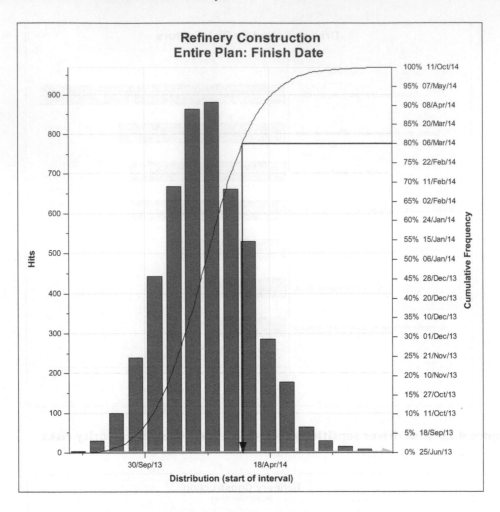

Figure 8.22 Schedule risk without Construction Supervision risk

To determine which risk should be next on the priority list for mitigation, a second tornado is produced. This is shown in Figure 8.23.

Notice two things about this tornado diagram:

- When the Construction Supervision Scarcity risk is eliminated the LLE Suppliers Busy risk jumps to the top (it was third in Figure 8.20). The list of priority risks has to be checked each time since risks' orders may change between one simulation and the next. This phenomenon is caused by the structure of the schedule itself. If you remove a strong construction risk from the equation the procurement risk may become more important since the procurement path is uncovered as the construction path becomes less risky. However, removing the procurement risk may not matter much if the construction risk is left unmitigated.

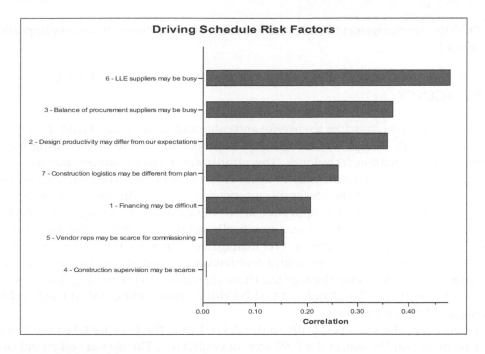

Figure 8.23 Sensitivity of schedule when Construction Supervision risk is eliminated[11]

- Figure 8.23 indicates that LLE Suppliers May be Busy' is the next most important risk. This may be so, but sensitivity is not the same as impact to the contingency at the P-80. Duration sensitivity concerns the correlation of risks with overall schedule durations which is calculated from the means. In our case, however, these refinery project stakeholders are more interested in the P-80 than the mean, so the priority order of risks to our stakeholders may differ from their order in the tornado diagram.

The next steps will be performed in series:

- take out (set probability to 0.0 percent) the next risk (at this point it is the 'LLE Suppliers May be Busy' risk);
- re-run the simulation, gathering the data as if the subject risk were 100 percent mitigated;
- examine the risk driver tornado chart to determine the next most important risk to mitigate;
- continue eliminating risks (repeat steps 1–3) until all risks are neutralized.

The remaining risk, in our case, is the background risk of -10 percent to +20 percent risk on the activity durations. If that were somehow to be mitigated we would be back to the deterministic date.

---

11   Note that the risk 'Construction Supervision may be Scarce' shows a zero correlation, since it has been assigned a probability of 0.0 percent.

Produce for management a table such as Table 8.3 that indicates the relative importance of the risks.

PRESENTING THE RISKS IN PRIORITY ORDER—EXPLAINING THE TIME CONTINGENCY

Using the risk drivers provides a powerful and calibrated priority list of risks. This priority list is a key result of sophisticated quantitative modeling of the project using familiar Monte Carlo simulation techniques. The prioritization takes complete account of the schedule, one of the main management documents of the project. The input data are based on the opinions of many knowledgeable interviewees. In these interviews the probability of risks' occurring and the multiplicative factor representing their ranges of impacts are all considered. The risks are prioritized specifically as they impact the schedule to the P-80 target confidence level or whatever level management chooses. Hence the prioritization is viewed as more accurate, informative and relevant to management's targets than the risks that are included in the risk register. These approaches to prioritizing risks have been evaluated recently in a handbook entitled *Prioritizing Project Risks, a Short Guide to Useful Techniques*. (Hopkinson 2008)

To prioritize the risks specifically to the desired date, the risks are taken out in order of their impact on the date at the P-80 level of confidence. The risks are taken out of the simulation in the sequence of those that make the largest improvement. Hence risks are listed in order of priority for management in Table 8.3.

Table 8.3   Prioritizing individual risks to the P-80 date

Prioritizing Schedule Risks using Risk Drivers			
Deterministic Date: 19-Jul-13	P-80	Impact on the P-80 Date	
Background plus All Risks	14-Apr-14	Days	% of Total
Risks Removed in Priority Order			
Construction Supervision Risk	06-Mar-14	39	14%
LLE Suppliers Busy Risk	29-Jan-14	36	13%
Balance of Procurement Suppliers Busy	17-Dec-13	43	16%
Construction Logistics Risk	05-Nov-13	42	16%
Design Productivity Risk	28-Sep-13	38	14%
Financing Availability Risk	18-Sep-13	10	4%
Vendor Representative Risk	09-Sep-13	9	3%
Background Risk	19-Jul-13	52	19%
Total		269	100%

Notice that the 'Days' saved and '% of Total' columns show some inversions between adjacent risks in that risks that are taken out down the line may seem to save more time than those that were taken out earlier. For instance removing the Balance of Procurement Suppliers risk saves 43 days while the LLE Suppliers Busy Risk taken out just above it saved only 36 days. This does not mean that the order should be reversed. Until the LLE

Suppliers Busy Risk is removed the Balance of Procurement Suppliers Busy risk is less important because of the structure of the schedule logic. The risks were carefully taken out in order of their marginal impact on the schedule.

Figure 8.24 overlays the results of taking out each risk in priority sequence as shown in Table 8.3. This figure indicates that the cumulative distribution shifts toward the origin and straightens up as more risks are removed in sequence. Thus the P-80 date is earlier and earlier as more risks are mitigated, and the resulting S-curve is straighter (the distribution is narrower), indicating more certainty in the answers. This chart is similar to the one independently constructed that appears in (Hopkinson 2008).

The priority table and chart are very useful for management to:

- explain how the contingency, in this case 269 calendar days, is generated;
- lead the risk mitigation effort in an efficient direction.

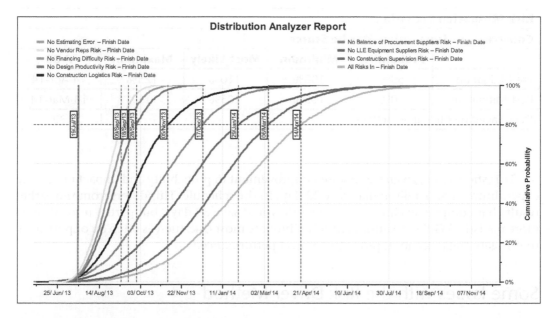

Figure 8.24 Effect on schedule risk as individual risks are removed in priority order

Analyzing Risk Mitigation

The mature project manager will look at these results and decide to mitigate risks rather than just accept the results as a foregone conclusion (the immature project manager may argue that the risk analysis results must be incorrect because the project cannot possibly be that late). The reasonable first step is to design a plan to mitigate the risk concerning scarcity of construction supervision. In days of high oil prices many oil companies were building refineries, chemical plants, LNG and GTL plants as well as upstream projects for extracting oil and gas. Each of these plants was predicated on being able to tap into the labor market for construction labor, including supervision. Add to this the construction boom in some parts of the world, the Middle East in particular, and the available pool of labor had shrunk. In particular, when the labor force on the project may be 20 000 workers, experienced and qualified supervision becomes a major concern.

The company may choose to spend $50 million with the constructors to ensure that high-quality professional and experienced supervisors will be assigned to this project rather than to the other projects the constructor is working on. How much will that help the schedule? We can develop a mitigation scenario that represents reductions in the probability that the construction supervision will affect this plant as well as the impact range of the risk should it occur. This scenario might be represented as shown in Table 8.4.

Table 8.4   Risk mitigation scenario specification and analysis of the results

Risk Mitigation Scenario					
Construction Supervision May be Scarce					
	Probability	**Minimum**	**Most Likely**	**Maximum**	**P-80**
Pre-Mitigation	50%	100%	110%	125%	14-Apr-14
Post Mitigation	20%	100%	105%	115%	13-Mar-14
Days Saved by Mitigation Actions					32

This shows that, given the assessed improvement in probability and impact range, the expenditure of $50 million may save 32 days in the schedule. This differs from the earlier result that completely eliminating this risk would save 39 days, but partial mitigation is more realistic. While we cannot entirely eliminate most risks there is often an opportunity to mitigate the risk and improve the project performance.[12]

Some Issues with the Risk Driver Method

There are some issues that need to be considered in using this method. For instance:

- With multiple risks affecting one activity, the risks might affect the duration in parallel or in series. That is, if two risks occur on the project, the project may recover from them simultaneously or have to recover from one and then the next in series. The Risk Factors module from Primavera Pertmaster assumes that the factors are multiplied, essentially forcing the risks to cumulate in series. This requirement that

---

12   We should note that time is money and if we feel we can save 32 days that could also reduce the amount of contingency reserve of money that would be needed. In fact, the reduction of the needed contingency of cost will partially make up for, or might even exceed, the $50 million that was required. These refineries are large multi-billion dollar projects so 32 days of construction, particularly at peak labor periods, could provide serious cost saving.

risk factors are multiplicative sometimes requires narrowing the risk factor ranges for risks that are assigned to activities that have several other risks already assigned.[13]

- The risk drivers are assumed to be independent whereas some risks might be correlated. This can be handled in some cases by consolidating risks that are deemed to be highly correlated, making the resulting risks in the risk list independent. However, if there are risks that cannot be consolidated, there may be a need to correlate the risks before applying them to the schedule. Working at the level of fundamental risks may lead to independence of those risks, but we cannot count on this.

Chapter Summary

This chapter introduces a new approach to schedule risk analysis, the risk driver method. The risk driver method is contrasted to the traditional three-point estimate schedule risk approach that:

- focuses on the impact of risks on activity durations by specifying three-point estimates of uncertainty in activity durations;
- does not allow specification of the level of probability of risks' occurring since the risk range is 100 percent likely. In the traditional approach we work only with a range of optimistic-most likely-pessimistic number of days' duration, not with the probability of the risks' occurring;
- can identify those activities and paths that determine the uncertainty in the result, but cannot identify the priority of the underlying risks.

The new Risk Driver approach:

- focuses on the risks themselves, usually benefiting from the risk register that prioritizes individual risks using qualitative methods. The link between qualitative and quantitative analysis is thus made explicit;
- characterizes the risks by both their probability of occurring and their impact on duration if they do occur. Some risks are more like ambiguities (for example, estimating error) or like uncertainties (for example, the degree of labor productivity). These types of risk can have probability of 100 percent while having uncertain impact ranges. Of course risk events will exhibit probability of occurring that is less than 100 percent;
- assigns the risks to all activities that they affect. A risk can affect multiple activities and an activity can be affected by multiple risks. Thus we explicitly model how risks affect the schedule;
- ensures that the activities whose durations are affected by common risks will be correlated. If one activity is only affected by the same risks as another activity their durations will be correlated 100 percent. Of course if two activities are affected by independent or confounding risks as well the degree of their correlation can be at any level (usually positive) less than 100 percent. The degree of correlation between

---

13   In some risk register applications (see Primavera Pertmaster) the user may specify parallel or series. The risk factors module does not at this writing (August 2008) allow this option.

activities is thus modeled, not assumed, and the correlation coefficient is the result of the model rather than 'guestimated' by interviewees;

- 'explains' the generation of a time contingency reserve needed by management by listing the risks in priority order and computing the days contributed by each so that management can judge importance of risks in determining the need for the contingency;
- prioritizes the risks rather than the activities and paths.
- This is excellent for the risk-mature organization that wants to focus on risk mitigation.

The risk driver approach promises to be an exciting and a more powerful approach to conducting schedule risk analyses than the standard three-point estimate approach traditionally used. Of course both approaches are based on solid CPM scheduling, thorough data collection and Monte Carlo simulation.

References

Hopkinson, M., et. al. (2008). *Prioritising Project Risks—A Short Guide to Useful Techniques*. Association for Project Management.

Hulett, D. and Whitehead, W. (2007). Using the Risk Register in Schedule Risk Analysis with Monte Carlo Simulation. *2nd Annual Oil & Gas Project Risk Management Conference*. Kuala Lumpur, Malaysia, October 29, 2007.

CHAPTER 9

Schedule Contingency Plans: Using Conditional Branching

Introduction

This book introduces the probabilistic way of thinking about schedules to assess schedule risk. The probabilistic approach allows more reality and flexibility in the schedule, mimicking what could happen in real projects. Conditional branching is perhaps the most powerful and sophisticated approach to making the schedule realistic by allowing changes in schedule logic, constraints and durations based on what happens in the project. The project is no longer a picture of a plan that is static as of a point in time. The project manager's discretion to change the plan as the schedule events dictate can now be represented by conditional branching.

Static schedules with their fixed logical relationships between activities are no longer a limiting factor when we need to represent project manager's options. Using conditional branching with schedule risk techniques and supporting software we can represent changes to the structure of the schedule itself that will occur if a schedule event occurs. One example of this capability that is explored in this chapter is the contingency plan in which an event, called a 'trigger' causes a pre-planned change in project strategy.

A trigger could be the late finish of design, which might cause a change in fabrication strategy. In each iteration of a Monte Carlo simulation the software tests the condition to see whether the trigger event has occurred. If the trigger has occurred, the software will make the specified changes to the schedule for that iteration. In some risk anlaysis software these changes may be to the logic, durations or constraints. If the trigger event has not occurred the software may make a different change, or no change, to the schedule.

Conditional branching is not probabilistic branching (see Chapter 7). Probabilistic branches occur with the toss of a coin, at random with some probability of occurring. Conditional branching is more focused and structured, since the change will be caused by a schedule event such as an activity or milestone being later than some date. Hence, conditional branching is deliberate and represents a conscious management decision. In contrast, probabilistic branching represents schedule events that happen at random.

Conditional Branching has Flexibility and Power

Events happen to projects that cause changes from the plan. Often the project manager will specify what trigger events could cause changes to the plan and what those changes would be. Unless the trigger events occur the current plan is followed. But, if some schedule event occurs the project manager plans to change direction, causing the very structure of the schedule to change by including new activities, excluding activities, changing successors, changing durations, accelerating or delaying activities or other changes. Conditional branching allows the scheduler to model those triggers and the resulting changes to the project schedule based on schedule events' occurring or not.

Triggers

Examples of schedule-based trigger events might include an activity taking longer than anticipated or finishing later than desirable. These trigger events often indicate problems with the project's basic assumptions and may indicate that the project objectives (time, cost, scope or quality) are in jeopardy. The project manager may decide that, under these circumstances, it would be better to change the original plan. Two examples from real projects illustrate some common trigger events:

• If a technology is more difficult than anticipated, leading to a longer than acceptable design phase, a backup or alternative technology may be adopted in order not to delay the project further. If the design period is not finished on a certain date, that condition may trigger a change in plans. This describes the typical contingency planning that occurs in many projects.
• The delay of fabrication of large modules that go on top of offshore oil or gas platforms may put some activities such as heavy lifting of that equipment onto an offshore oil or gas production platform into the period of heavy weather (for example, monsoons, winter), making those activities dangerous, take longer or cost more than anticipated. Hence a fabrication delay may be the trigger that causes heavy lifting to be delayed until after the heavy weather.

The essence of these conditions or triggers is that they are schedule events with important consequences. To use them as conditions or triggers they must be based on some schedule phenomenon such as a date or duration. The trigger event will either occur or not in an iteration of Monte Carlo simulation because of uncertainty in an activity's durations or those of its predecessors.

Consequences

Conditional branching allows the trigger event to cause many different types of changes in the affected activities. Some examples of changes that are available include:

• Change the duration of activities. Some activity duration may be changed based on a condition. A dramatic example of a duration change might be to make the duration zero days, effectively eliminating that activity from the project schedule in favor of another activity. Another would be to increase the duration of the activity. For instance, if the schedule puts structural steel erection or concrete pouring in the winter instead of the summer, the durations of those tasks might increase.

- Change a successor activity. This might lead to a different path becoming critical if the trigger were to occur. Here, for example, an alternative technology might be chosen.
- Change a constraint. An activity that starts 'as-soon-as-possible' (ASAP) in the base schedule could 'start-not-earlier-than' (SNET) some date, say after the winter turns to spring, if the trigger occurs.

The Structure of a Conditional Branching Statement

The conditional statement is structured as follows:

If *(condition occurs)* then *(alternative 1)* else *(alternative 2)*

A typical use of conditional branching starts with a condition such as a particular activity taking longer than anticipated or finishing later than planned. Taking longer or finishing later may indicate that the project is in trouble and may activate an alternative, call it Plan B, to the original plan.

The conditional branch will represent the contingency that has been planned to take either one of two (or more) alternative courses of action based on an event's happening. Activities representing these courses of action must all be present in the schedule. The essence of the conditional approach is that these actions are alternatives to one another—they will not both exist in any one project.

Two types of conditional branching will help explain the capability:

1. A slip of an activity beyond a specific date represents failure or unreasonably jeopardizes project objectives such as schedule, scope or cost. The decision may be to abandon the preferred Plan A and adopt instead a contingency plan, Plan B. This is the typical contingency planning situation found in most descriptions of project management. This use of conditional branching is shown in the next section.
2. The schedule event may be determined by something physical such as the height of the sea swells during monsoons or winter that jeopardize setting of an offshore production platform. Any of these conditions makes lifting of expensive (for example, $100 million), heavy (for example, two thousand tons) production modules activities inadvisable, because they could be damaged or lost. Any weather-related activity pushed past a specific date has to wait until the end of weather period to start. An example of this is in the second section below.

Policy Alternative: Trade-off Between a Preferred but Risky Approach versus Timely Completion

To illustrate conditional branching, consider this scenario. A government agency wants to develop a warship with many new capabilities. In particular, a new and technically advanced engine (preferred Plan A) for a new generation warship is included in the project plan. The engine is to be powerful, quiet, energy efficient and produce low emissions into the atmosphere. In other words, this is a dramatically new engine with many potential problems in development that could take time to resolve.

The other concern is schedule. The customer does not want to slip the launch date. They are even willing to forego the new propulsion system, if that becomes necessary, to preserve schedule for the first such warship. The customer asks the contractor's project management team to look at installing a conventional propulsion system (Conventional Backup Alternative Plan B in Figure 9.1) for the first article as an alternative to keep the schedule under control.

The schedule shown in Figure 9.1 is a summary of the propulsion development that becomes part of the project schedule. It shows Design, Build and Test of the new propulsion system as New Preferred Alternative A (Plan A) and the conventional system as Conventional Backup Alternative B (Plan B). Plan B is only going to be used if Plan A is judged to be in trouble. This may occur, for example, if the technology design is more difficult than anticipated. However, to be a viable contingency plan, Alternative B must at least be designed for this ship's hull and be ready for Build and Test (Activity Design Alt B in the schedule Figure 9.1).

In order to develop a strategy for completing this project on or as close to January 24, 2011 as possible, we first determine the risks for each of the four activities. From the schedule we can see that Alternative A takes 800 days and Alternative B takes 760 if everything goes according to plan. During interviews of the project participants we determine that the risk for the two activities of Alternative A was greater than the risk on the corresponding Alternative B activities. This is logical since Alternative A is uncharted waters and Alternative B is mainly application of known, conventional technology to a new ship's hull. These findings are incorporated in the three-point estimates shown in Figure 9.2.

First we probably want to know whether there is even a need for a contingency Plan B. To find out if Plan B is needed we conduct a schedule risk analysis only on Plan A. The results are shown below in Figure 9.3.

The customer sees that the schedule without the availability of any backup technology at the 80th percentile (the P-80) is September 13, 2011 compared to the scheduled date of January 21, 2011. This 8-month slip in schedule is unacceptable and they ask for alternatives. To help answer the question: 'What good will a backup Plan B be for our schedule?' we need to see how much schedule delay there would be if we had a backup Plan B using conventional propulsion.

ID	Task Name	Duration	Start	Finish	2008				2009				2010				2011	
					Qtr 4	Qtr 1	Qtr 2	Qtr 3	Qtr 4	Qtr 1	Qtr 2	Qtr 3	Qtr 4	Qtr 1	Qtr 2	Qtr 3	Qtr 4	Qtr 1	Qtr 2
1	Start	0 d	1/1/08	1/1/08		1/1													
2	New Preferred Alternative A	800 d	1/1/08	1/24/11															
3	Design Alt. A	300 d	1/1/08	2/23/09		1/1				2/23									
4	Build and Test Alt. A	500 d	2/24/09	1/24/11						2/24							1/24		
5	Conventional Backup Alternative B	760 d	1/1/08	11/29/10															
6	Design Alt. B	285 d	1/1/08	2/2/09		1/1				2/2									
7	Build and Test Alt. B	475 d	2/3/09	11/29/10						2/3						11/29			
8	Finish	0 d	1/24/11	1/24/11													1/24		

Figure 9.1   New ship propulsion system summary schedule showing preferred Plan A and backup Plan B

ID	Task Name	Rept IC	Min Rdur	ML Rdur	Max Rdur	Curve
1	Start	0	0 d	0 d	0 d	0
2	**Preferred Alt A**	**0**	**0 d**	**0 d**	**0 d**	**0**
3	Design Alt. A	0	250 d	300 d	500 d	2
4	Build and Test Alt. A	0	450 d	500 d	700 d	2
5	**Backup Alt. B**	**0**	**0 d**	**0 d**	**0 d**	**0**
6	Design Alt. B	0	250 d	285 d	325 d	2
7	Build and Test Alt. B	0	425 d	475 d	525 d	2
8	Finish	2	0 d	0 d	0 d	0

Figure 9.2   Risk ranges on Alternative A and Alternative B activities show that Alternative A has more schedule risk

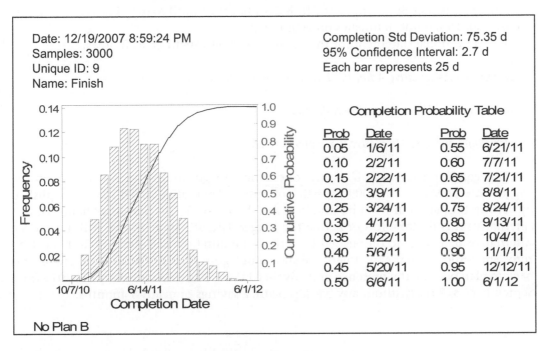

Figure 9.3   Without Plan B the 80ᵗʰ percentile is September 13, 2011

Contingency planning involves several steps:

- identify the part of the project plan, called Plan A, that may need a contingency plan;
- determine the alternative to that plan, called Plan B, that will be activated if Plan A is deemed to be in such trouble that it must be abandoned;
- specify the definition of the point at which Plan B will be chosen if something happens or does not happen. This is often called a Trigger. If the trigger occurs in any iteration, Plan A will be dropped and Plan B will be initiated.

In this case we determine that the initial trigger is whether the design of Plan A can be completed in a timely fashion. That date in the schedule is February 23, 2009 but we give it until March 10, 2009 to provide for some acceptable slippage.

The way this conditional branch works is that if Design Alternative A is not completed by March 10, 2009 then Plan A is deemed to be in trouble. Plan B, the conventional engine, must be adopted for the first ship produced if Design Alternative A finishes later than that date.

Before we can implement this contingency plan we need to make one more adjustment. Build and Test Alternative B must not start before March 11, 2009, the morning after the decision might be made to abandon the new engine for the first ship. Clearly, even if Design Alternative B is complete before that date, the project will wait until the decision has been taken before beginning the expensive tasks to build and test the Alternative B engine. Hence, we put a 'start-no-earlier-than' (SNET) constraint of March 11, 2009 on Activity 7, Build and Test Alternative B. The result of setting this constraint, shown in Figure 9.4, is that Build and Test Alt. B does not start until March 11 even though its predecessor Design Alt. B finishes on February 2.

The conditional statement for this example can be structured as follows:

If *(Design of Alternative A finishes later than March 10, 2009)*

then *(Build and Test Alternative A duration = 0 days)*

else *(Build and Test Alternative B duration = 0 days)*

This condition says that if, on the one hand, Design Alternative A is later than the trigger date of March 10, 2009 we will not pursue Alternative A. This is accomplished by setting the duration of Build and Test Alt. A to zero. If, on the other hand, Design Alternative A is on time (on or before the trigger date of March 10) we will not pursue Alternative B. This is accomplished by setting the duration of Build and Test Alt. B to zero. The results of that experiment are shown in Figure 9.5. The customer approves this backup plan because the presence of it advanced the P-80 date to February 25, 2011 from September 13, 2011 (without any backup plan), a savings of nearly 7 months.[1]

ID	Task Name	Duration	Start	Finish
1	Start	0 d	1/1/08	1/1/08
2	**New Preferred Alternative A**	**800 d**	**1/1/08**	**1/24/11**
3	Design Alt. A	300 d	1/1/08	2/23/09
4	Build and Test Alt. A	500 d	2/24/09	1/24/11
5	**Conventional Backup Alternative B**	**786 d**	**1/1/08**	**1/4/11**
6	Design Alt. B	285 d	1/1/08	2/2/09
7	Build and Test Alt. B	475 d	3/11/09	1/4/11
8	Finish	0 d	1/24/11	1/24/11

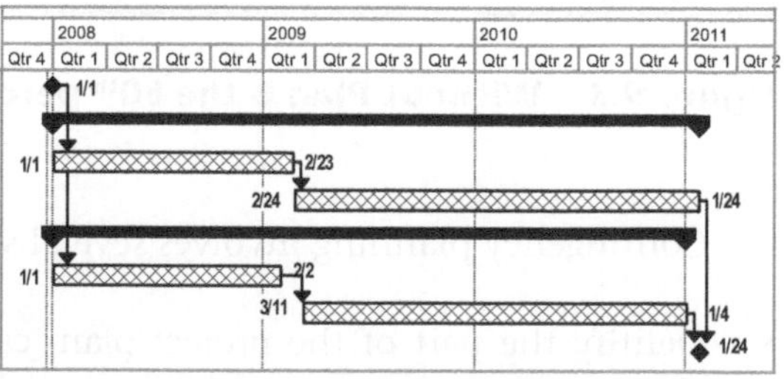

Figure 9.4 Schedule with Build and Test Alternative B set with a start-no-earlier-than constraint of March 11, 2009

1 This example uses Risk+ from Deltek. Risk+ has a particularly flexible conditional branching capability that is implemented in Visual Basic Applications (VBA).

The comparison of the S-curves from the two schedule risk results is shown in Figure 9.6.

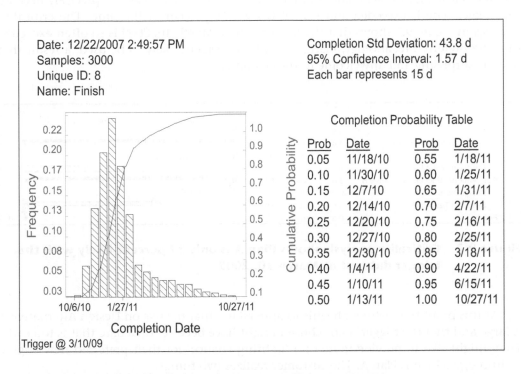

Figure 9.5   Schedule risk with trigger set at March 10, 2009

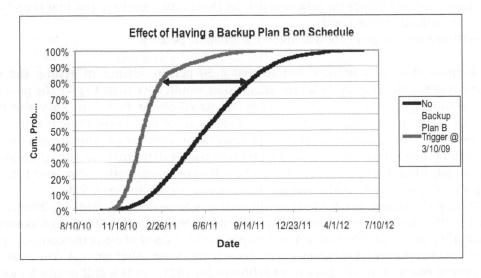

Figure 9.6   Schedule risk effect of a backup Plan B with a trigger set at March 10, 2009

In spite of the pleasure the customer feels at having saved nearly 7 months there may be a slight question about how that saving was accomplished. Upon further analysis it becomes clear that the schedule is much shorter with a backup Plan B precisely because the trigger date is exceeded and Plan B is used 68 percent of the time. The criticality analysis in Figure 9.7 shows that the trigger date of March 10, 2009 is so often exceeded that the preferred technology, Plan A, is only 32 percent likely to be used, a result that may be surprising and unwelcome to the customer.

ID	Task Name	Total Slack	Critical	% Critical	2008				2009				2010				2011	
					Qtr 4	Qtr 1	Qtr 2	Qtr 3	Qtr 4	Qtr 1	Qtr 2	Qtr 3	Qtr 4	Qtr 1	Qtr 2	Qtr 3	Qtr 4	Qtr 1	Qtr 2
1	Start	0 d	Yes	34	12/22													
2	New Preferred Alternative A	0 d	Yes	32														
3	Design Alt. A	0 d	Yes	32	2/25				2/13									
4	Build and Test Alt. A	0 d	Yes	32					2/16							1/14		
5	Conventional Backup Alternative B	8 d	No	68														
6	Design Alt. B	40 d	No	2	2													
7	Build and Test Alt. B	8 d	No	68	68													
8	Finish	0 d	Yes	100												1/14		

Figure 9.7 **Criticality analysis shows Plan A is only 32 percent likely with the trigger date set at March 10, 2009**

At this point the customer begins to understand that the risk on Design Alternative A is large and that the trigger point chosen might have been set on a date that is too early. They might like to provide more of a 'fighting chance' for their preferred new engine technology which is Plan A. The customer realizes two things:

1. to achieve more of a chance of getting the preferred new engine we need to relax the trigger date (or reduce the schedule risk on Design Alternative A, but that is probably more difficult at this point);
2. setting a later trigger date will involve a schedule penalty.

Suppose that the customer decides that a 50 percent chance of getting the new engine is desirable. They ask what the trigger date would have to be to give the preferred new technology a 50 percent chance of being placed on the first ship. They also need to know what the schedule penalty at the P-80 is, since they know there is a trade off between these two objectives. A little experimentation with various simulations and the conditional branching software, trying different trigger dates that are later than March 10 (and adjusting the SNET constraint on Plan B to be consistently 1 day later than the trigger date) yield the desired result and implications shown in Figure 9.8.

Setting the trigger date for Design Alternative A to be done by April 15, 2009, 2 full months after that activity's scheduled completion date, results in a 50 percent likelihood of installing preferred Alternative A and 50 percent likelihood of using the contingency B given the uncertainty on the duration of the activity Design Alternative A. The P-80 date has moved out to April 18, 2011 from February 25, 2011, but it is still nearly 5 months earlier than September 13, 2011 that would occur at the P-80 without a backup at all. The three alternatives examined are compared in Figure 9.9 and Table 9.1.

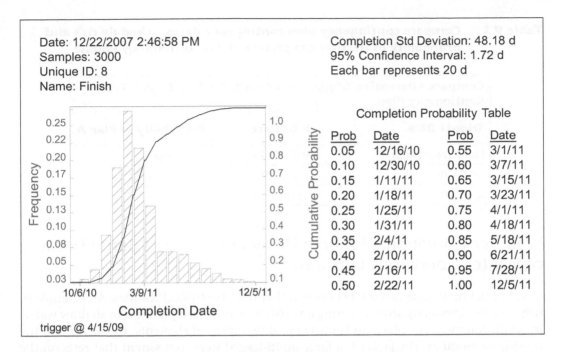

Date: 12/22/2007 2:46:58 PM
Samples: 3000
Unique ID: 8
Name: Finish

Completion Std Deviation: 48.18 d
95% Confidence Interval: 1.72 d
Each bar represents 20 d

Completion Probability Table

Prob	Date	Prob	Date
0.05	12/16/10	0.55	3/1/11
0.10	12/30/10	0.60	3/7/11
0.15	1/11/11	0.65	3/15/11
0.20	1/18/11	0.70	3/23/11
0.25	1/25/11	0.75	4/1/11
0.30	1/31/11	0.80	4/18/11
0.35	2/4/11	0.85	5/18/11
0.40	2/10/11	0.90	6/21/11
0.45	2/16/11	0.95	7/28/11
0.50	2/22/11	1.00	12/5/11

trigger @ 4/15/09

Figure 9.8   Schedule risk results with trigger set at April 15, 2009

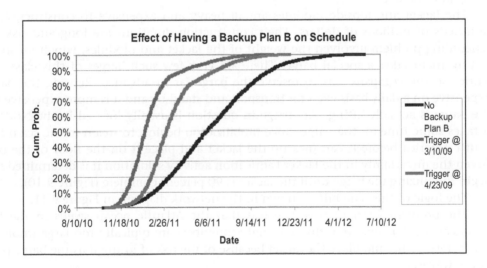

Figure 9.9   Comparing no backup, early trigger and final 50-50 trigger schedules

The alternative trigger dates with their implications for completion (at the P-80) and chance of using the customer's desired motor, are shown in Table 9.1.

Table 9.1 **Compare contingency plan contingency dates, schedule risk and probability of getting the preferred new technology engine**

Compare Alternative Triggers and Results for Ship Engine Type Contingency Plan		
Trigger Date	**P-80 Date**	**Probability of Plan A**
None (No Plan B)	9/13/2011	100 percent
3/10/2009	2/25/2011	28 percent
4/15/2009	4/18/2011	50 percent

Weather Conditional Branch: Delay of Heavy Lift Barge because of Winter Conditions Offshore

Conditional branching can set a condition that reflects physical realities. An example of this was the scheduling around setting an offshore production platform in shallow waters in South America. The platform is made up of two types of elements, the jacket and the topside or modules. The jacket is a large multi-legged steel component that rests on the ocean floor on top of which are placed the modules for pumping, electricity generating, living and the like.

The jacket and topside modules are all heavy and expensive to construct, costing hundreds of millions of dollars. Their construction durations are long and risky. The scheduling problem involved the weight of the jacket and modules, which means that setting them takes a special barge. There are very few such barges of this class in the world, so the company has to reserve the barge far in advance. Renting the barge is expensive on a daily basis even for transport and stand-by and it is more expensive when the barge is actually lifting. The barge in question is usually located in the North Sea. It takes some time for the barge, once mobilization begins, to steam to the sea offshore South America before it can pick up the jacket and place it on the floor of the ocean. Given the uncertainty in the Jacket Fabrication activity's duration it is determined not to begin mobilizing the barge until the jacket is 90 percent complete (Figure 9.10).

The logic of this schedule is shown in the network diagram in Figure 9.11.

The problem faced in this case was that the Atlantic ocean swells in the South American winter (mid-May through mid-September) are typically too large (more than 1.5 meters) to lift and place the jacket because of the risk of losing it as the barge pitches in the waves.

The project manager is pleased that the schedule shows that setting of the jacket will be complete by April 25, 3 weeks before the winter season. Looking at the deterministic schedule the customer believes that the jacket will be set before the southern winter. Of course, by now the project manager knows that there is an uncertainty in the jacket construction, barge mobilization and setting the jacket activities. These uncertainties are shown in Table 9.2.

Given these risks, a Monte Carlo simulation shows the probability distribution for the Finish date (Figure 9.12).

ID	Task Name	Duration	Start	Finish
1	Start	0 d	2/1/08	2/1/08
2	Jacket Fabrication to 90%	400 d	2/1/08	3/6/09
3	Complete Fab. and Transport Jacket	40 d	3/7/09	4/15/09
4	Module Construction	400 d	2/1/08	3/6/09
5	Transport Modules	45 d	3/7/09	4/20/09
6	Mobilize the Barge	40 d	3/7/09	4/15/09
7	Set the Jacket	10 d	4/16/09	4/25/09
8	Install the Modules	50 d	4/26/09	6/14/09
9	Finish	0 d	6/14/09	6/14/09

Figure 9.10 Base schedule showing the jacket safely set by April 25

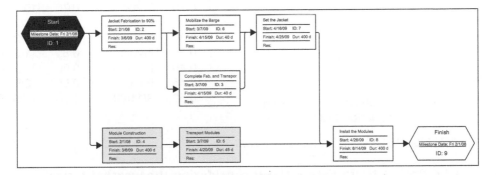

Figure 9.11 Network diagram showing the logic in the base schedule

Table 9.2   Risk ranges on the activity durations for the base schedule

ID	Task Name	Rept ID	Mn Rdur	ML Rdur	Max Rdur	Curve
1	Start	0	0 d	0 d	0 d	0
2	Jacket Fabrication to 90%	0	350 d	400 d	600 d	2
3	Complete Fab. and Transport Jacket	0	35 d	40 d	55 d	2
4	Module Construction	0	350 d	400 d	600 d	2
5	Transport Modules	0	30 d	45 d	75 d	2
6	Mobilize the Barge	0	35 d	40 d	60 d	2
7	Set the Jacket	1	8 d	10 d	20 d	2
8	Install the Modules	0	40 d	50 d	75 d	2
9	Finish	2	0 d	0 d	0 d	0

While these dates do not give June 14, 2009 much of a chance (it is less than 5 percent likely) the P-80 date of October 29, 2009 is only about a 4 month slip. This schedule delay, providing for an 80 percent chance of success, is deemed to be acceptable to the customer.

Someone notices that there may be a weather problem that has not been figured into the analysis. If setting the jacket occurs during the winter the schedule results shown above will be adversely affected. (The same person notices that lifting and setting the modules appears squarely in the winter even in the base schedule.) The project manager looks at the results for the activity # 7, Set the Jacket, and finds that it indeed is very likely to be conducted in the worst part of winter in the southern seas. Figure 9.13 indicates that there is a preponderance of dates in the period of high seas, when it makes no sense to try to set a large, heavy and expensive piece of equipment on the floor of the ocean.

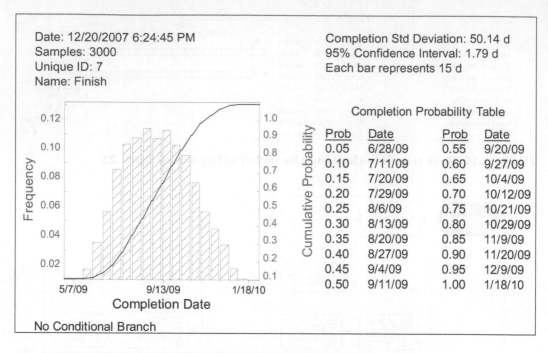

Figure 9.12 Project finish date without considering the weather factor

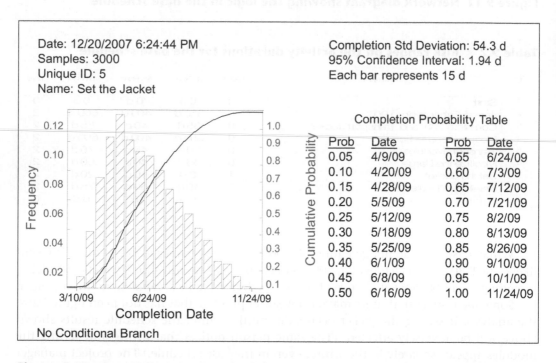

Figure 9.13 Uncertainty in the schedule puts setting the jacket in the southern winter

Seeing this result the project manager has to reconsider the viability of the base schedule. When risk is considered the jacket is not very likely to be set before the seas become inhospitable. Even the risk analysis on the base schedule without contingency is invalidated by the weather consideration.

The project manager decides to determine a 'drop dead' date for Jacket Fabrication to 90 percent of March 26, 2009. After that point, they decide, it would be too late to call on the barge for the season. If the project misses the season for heavy lifts it would be better to delay the beginning of barge mobilization until August 15, targeting its arrival for after the end of the southern hemisphere winter. The project manager decides to use conditional branching with the conditional statement[2]:

If *(Jacket Fabrication to 90 percent finishes later than March 26, 2009)*

then *(the start date for Mobilize the Barge is set at August 15, 2009)*

else *(Mobilize the Barge starts as-soon-as-possible (ASAP)).*

Setting this condition the Monte Carlo simulation is re-run with the following results for the Finish task (Figure 9.14).

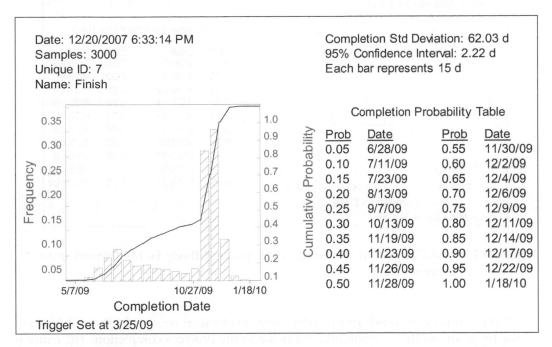

Date: 12/20/2007 6:33:14 PM
Samples: 3000
Unique ID: 7
Name: Finish

Completion Std Deviation: 62.03 d
95% Confidence Interval: 2.22 d
Each bar represents 15 d

Completion Probability Table

Prob	Date	Prob	Date
0.05	6/28/09	0.55	11/30/09
0.10	7/11/09	0.60	12/2/09
0.15	7/23/09	0.65	12/4/09
0.20	8/13/09	0.70	12/6/09
0.25	9/7/09	0.75	12/9/09
0.30	10/13/09	0.80	12/11/09
0.35	11/19/09	0.85	12/14/09
0.40	11/23/09	0.90	12/17/09
0.45	11/26/09	0.95	12/22/09
0.50	11/28/09	1.00	1/18/10

Trigger Set at 3/25/09

Figure 9.14 Project completion when setting the jacket does not occur during winter

---

2   It would be tempting to use a resource applied to the Jacket Setting activity, where the resource, call it Weather, would have a calendar with the winter dates designated as non-working. While this would handle the actual setting of the jacket it would not handle the barge problem. Using this approach the barge would be mobilized based on the risk in the schedule for 90 percent fabrication of the jacket but would, probably, arrive on location and stand by waiting to be able to lift the jacket. At a rental cost of US 350 000 per day, this would quickly become a losing strategy.

The project manager examines this strange-looking probability distribution and sees that the P-80 date is now December 11, 2009 instead of October 29, 2009 adding a month and a half to the completion date. They wonder what happened to cause this slip in the results.

The conditional branch was set up on the Barge Mobilization activity. Looking at the Mobilization of the Barge the project manager finds that the schedule is now more realistic given the uncertainties and the winter weather condition, but also that there is a smaller likelihood that the jacket will be set in the summer 2008–2009 season. The probability is about 65 percent that setting the jacket will miss that season and have to wait for the end of bad weather to set the valuable jacket on the ocean floor, as shown in Figure 9.15.

The two results are compared in Figure 9.16. The impact of the winter weather on the setting of the jacket would be more pronounced if the customer were to make its decisions on the 50th percentile rather than the 80th percentile.

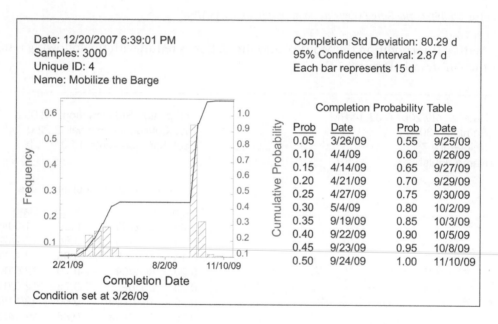

Figure 9.15 Mobilization of the barge is 65 percent likely to be delayed until after winter

Using conditional branching to model the effect of winter weather for the mobilization of the barge shows the true potential for delay to the project's completion. The cause is the interaction of the risk in Jacket Fabrication to 90 percent with the limitation imposed on Set the Jacket by the high seas from May through September. Since the weather is a risk that is not manageable, the only action that will allow setting the jacket before winter is to mitigate the risk in fabrication to 90 percent. At least the project manager now has a realistic plan and completion date expectations that account for the winter weather problem.

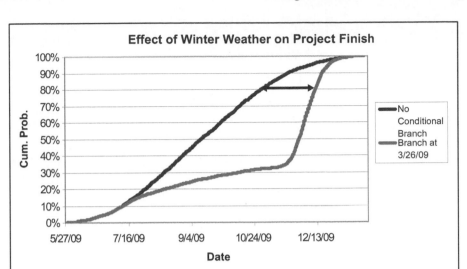

Figure 9.16 Comparing cumulative distributions with and without the winter weather condition

Summary of Conditional Branching

Conditional branching causes the schedule to change based on a condition that can be based on a schedule event, such as an activity being later or taking longer than planned. Whether the condition occurs in any iteration of a Monte Carlo simulation depends on the risks in the schedule including risks of predecessor activities. If the condition (the trigger event) occurs, then durations, logic or constraints may be altered.

The consequences can be flexible. If the condition occurs, activities can appear or disappear; logical relationships can change; constraints can appear, disappear or change; activity duration risks can increase or decrease—anything that one can program in Visual Basic Application (VBA) macros may be available to use, depending on the schedule risk software used.

Because we are modeling a schedule that can change with schedule risk, conditional branching is the most sophisticated modeling that can be done in schedule risk analysis today. It is intended to mirror the most important factor, the project manager's discretion in the face of schedule events. Unlike traditional deterministic scheduling where all activities in the schedule occur with fixed logic ties, durations and plan logic is static, conditional branching allows us to represent the project manager's decision making as it may be driven by schedule events.

Conditional branching is cause-and-effect driven, not simply random as in probabilistic branching. The method is implemented in different ways by different simulation software; the most powerful and flexible approach uses VBA. While VBA may be daunting to some people, the software vendors that have this capability include templates or examples to help people who are not VBA experts specify even complex 'if ... then ... else' logic. The main issue involves determining how to represent the condition and its alternatives logically. While this modeling takes care, conditional branching is rewarding since it represents the discretion available to project managers.

10 *When Activity Durations Move Together: Incorporating Correlation*

Introduction

In schedule risk analysis we do not believe that all activities will be at their most pessimistic or most optimistic durations in the same project. But, what would happen if many activities were to overrun their estimated durations together on our project? That event could cause very large overruns of the total project completion date. This is because the long durations are summed down the paths, reinforcing overruns of other long-duration activities on the same path. We can handle this common occurrence in probabilistic analysis using correlation between uncertain activity durations.

Correlation between activity durations is when activities systematically overrun (or under-run) their estimated durations *together* on a project:

- Is correlation between activity durations possible on my project? Yes, correlation is fairly common in project management.
- What would cause activities to overrun or under-run their durations together? Activities being affected by the same risks would be correlated.
- What would be the impact of strong correlation of durations on my target percentile, say the P-80? The P-80 date would be later than if there were no correlation simply because correlated durations reinforce each other as dates are accumulated down the paths.

In earlier chapters we assumed that activity durations were independent of each other, so there is a fair amount of canceling out between long, short and moderate durations down schedule paths. Offsetting long activities with short activities, which occurs with 'independence' or the absence of correlation, moderates the effect of risk on completion dates.

This chapter examines the assumption of independence between activity durations and introduces correlation. It turns out that correlation affects the results, particularly the upper and lower percentile results.

Correlation Concepts

Implementing correlation between the durations of activities that are affected by the same risks ensures that, during simulation, we do not have any iteration that represents an inconsistent scenario. Suppose that we may have trouble implementing a technology and we believe that if that happened we would have trouble with design, fabrication and testing of the product. During simulation we should not produce any iteration (scenario) that assumes the design of a product much longer than anticipated but the fabrication or testing much shorter than anticipated. This would be an inconsistent scenario. In reality we believe the three activities would all be influenced by risk in the same direction in any project so durations that go different directions from their means would be inconsistent with our assumption that they move together.

The problem of creating inconsistent scenarios during simulation arises from a purely random selection of durations. Selecting the durations at random will probably yield some iterations in which one or another of these activities had durations that were longer than their means while the others are shorter. In other words, a purely random selection of durations from input probability distributions would produce inconsistent scenarios. Examining and implementing correlation is intended to ensure that each iteration specifies an internally consistent scenario.

Correlation is caused by a risk affecting two activities on the same project. An example would be the same Risk Event affecting the duration of software coding and testing. If the risk occurs on a project and software coding is longer than scheduled, it stands to reason that software testing, affected by the same risk, would also be longer than scheduled. These activity durations exhibit correlation. Examples of such risks and their impact on multiple activities could include:

- poor labor productivity would affect the duration of all construction activities in the same project;
- technology difficulties would affect many design, fabrication and testing activities similarly;
- difficulty in making management decisions would lengthen all decision-making activities from their scheduled durations;
- if management imposed optimistic bias on the schedule, for instance to make it appear to be shorter, most, if not many, of the activities that are shortened might overrun their scheduled durations.

The mechanism of correlation is shown in Figure 10.1.

Correlation does not imply causality between the activity durations themselves, just between the risk and the activity durations. The activity durations move as if one is driving the other, but that mechanism is not needed with correlation. If one activity's duration causes the duration of another to vary, the relationship is causality and can be modeled. For instance, suppose fabrication of a procured item drives the level-of-effort procurement management activity. Procurement management may be represented by a hammock or summary activity which takes its duration from the duration of the fabrication activity. The duration of the detailed fabrication activity and its hammock exhibit perfect correlation because they will both be long or short together. (If there

Figure 10.1 One risk driving two activities causing them to be correlated

are several activities in the hammock the direction of causality will be the same but the degree of correlation may not be perfect.) However, activities and the hammocks that they drive are causally related and correlation need not be applied.

Correlation is defined between two activities. While many activities may be correlated, the correlation is defined between them in pair-wise fashion.

Activities that are correlated move together but not necessarily in lock step. Some correlated activities will move very closely together while others will just tend generally to move together but be influenced by other factors. The degree of correlation is measured by a 'correlation coefficient.' The correlation coefficient can be any value between -1.0 and +1.0.

- *Positive correlation* exists if, when one activity is above its expected value the other is predictably above its own expected value.[1] Activity durations that are positively correlated can be expected to be long or short together on any project, probably because they are influenced by a common risk or set of risks. Positive correlation is fairly common in project risk management. These correlation coefficients range between zero and one (+1.0).
- *Negative correlation* exists if, when one activity is above its expected value the other is predictably below its own expected value. Negative correlation is not common in project management. Negative correlation may occur when scope of work is re-assigned from one activity to another; if the work is moved from one activity its duration will be shorter than expected and the receiving activity will take longer. These correlation coefficients range between zero and minus one (-1.0).
- *Independence* is present when the duration of one activity within its probability distribution is not related to that of another activity. Independent activities exhibit a correlation coefficient of zero. We have assumed independence in earlier chapters in this book.

The strength of correlation between activity durations is measured by the size of the correlation coefficient. Taking positive correlation, the closer to a coefficient of + 1.0 the

---

1   Each of these activities will be varying within their own probability distribution so the number of days that one activity exceeds its expected value is generally not the same as that for the other correlated activity of the pair, even if correlation is perfect.

closer the correlation is to perfect correlation between the two activities and the more predictable it is that their durations will 'move together.'

Two activities' durations can be found to be correlated because common risks affect them. There are three conditions that affect the degree of correlation between different activities' durations:

- The force that drives the durations must be common to both activities. For example, if a risk occurs on a project it might influence the duration of (at least) two activities in the same duration and, presumably, with similar magnitudes.
- The force that drives the durations must be influential in determining activities' durations. For instance, if a risk affects strongly one activity's duration but only weakly affects another activity's duration, correlation may be present but will not be strong.
- The force that drives the durations are variable in the project. Risks have variability in their probability of occurrence and impact if they do occur, so risks are common causes of correlation.

What determines the strength of correlation between the durations of two activities? The degree of correlation depends on the risks that are present that affect the durations of two activities and other risks that may be present but are not common to the two activities. Take the situation where there is one risk that is common to two activities and other risks are absent. This situation causes perfect correlation (within rounding error) between the durations of a pair of activities. This situation is shown in Figure 10.2.[2]

The degree of correlation is reduced if there are other confounding risks that affect the activities but are not common to them. Figure 10.3 shows that the correlation is reduced to .48 when two confounding (non-common to the two activities) risks are introduced.

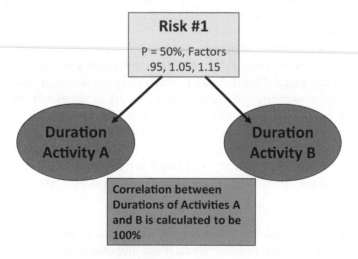

Figure 10.2 Perfect positive correlation

---

2   In these examples the risk was assigned and a simulation of 1000 iterations was run. The resulting 1000 pairs of durations were saved to Microsoft Excel® and its correlation function calculated the correlation between the two durations.

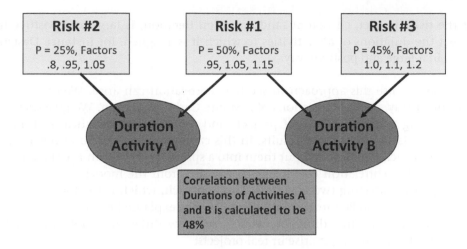

Figure 10.3 The presence of non-common or confounding risks reduces the degree of correlation

Specifying Correlation between Activity Durations

The traditional approach to specifying activity duration risk is to estimate the three-points of optimistic, most likely and pessimistic duration and to choose a distribution type (triangular, Beta, uniform and normal are common). If the durations of two activities are correlated, the correlation coefficient must be determined by examining history or making judgmental estimates.[3] After the pair of correlated activities is identified we need to decide on an estimate of the correlation coefficient (from -1.0 to +1.0 but generally between 0.0 and +1.0). Correlation coefficients are then assigned to the pair.

For those who are unfamiliar with correlation it might be useful to collect data on correlated durations and compute the correlation coefficients. Then, when faced with a specific project, the team can refer back to this exercise to estimate the coefficients used in the simulation. Unfortunately we don't usually have historical data from which we can compute actual correlation coefficients.

As an alternative to examining databases, which might not exist, the newer risk driver approach to schedule risk analysis (see Chapter 8) allows us to model the correlation between activity durations. With risk drivers the risks are identified and their parameters calibrated, then they are assigned to activities. Correlation between activity durations arise naturally when the same risk is assigned to two different activities. In that approach a risk has a probability of occurring and an impact on duration of two (or several) activities if it occurs. It would be common practice to assign this risk to several activities, for example, Design and Test. Then, during a project:

•   If the risk occurs in any iteration it has an impact on both the Design and Test activities' durations. If the risk does not occur in any iteration, neither Design nor Test activities are affected, at least by this risk.

---

3   In the next section we discuss how difficult choosing the coefficient may be.

- If the risk's impact, chosen at random for an iteration, is large and positive it will affect the durations of all activities to which it is assigned, for instance Design and Test, in a large and positive way.

It is easy to use this approach to see how correlation can arise. We can generate a large number (say, 5000) of iterations of a Monte Carlo simulation. We take their results as representing a large database of projects and use standard statistical techniques to compute correlation from those results. In this case we collect the durations for Design and Test from each iteration and put them into a spreadsheet. Within the spreadsheet we can calculate the correlation coefficient that results from the modeling of the external technology risk's affecting two activities. This approach, which is the foundation of the results shown above in Figure 10.2 and Figure 10.3, is explained in more detail.

We can examine three different cases to see how different correlation coefficient values, from 0.0 to 1.0, might arise in real projects:

- In one case we use simple unrelated duration risk based on traditional three-point estimates. If the risks are random and not related, then the computed correlation coefficient should be zero.
- In a second case we use the risk drivers approach, introduced in Chapter 8, to model how correlation arises. That is, correlation is caused a single risk affecting two different activities' durations—hence their durations become correlated approximately 100 percent (see Figure 10.2).
- Finally, we assign one risk to two activities as before, but in this case each activity also has a different risk. In this case the correlation between activities exists, because of the common risk, but is less than 100 percent because of the two independent risks also at work (see Figure 10.3).

Figure 10.4 shows just two activities, Design and Test. Design is set for 100 days and Test is 50 days, illustrating that correlation does not mean the same absolute number of days but rather the degree of agreement in the way the days move in terms of direction and magnitude within their own distribution across projects.

INDEPENDENCE

First we generate 5000 projects, represented by the 5000 iterations of a Monte Carlo simulation, using uncorrelated random variables. To generate uncorrelated durations we use the standard three-point estimate to create uncertainty in duration for two activities, Design and Test. We have assumed that each of these activities has a range of -10 percent to +20 percent and have not imposed any correlation. If these two activities are not

ID	Description	Jan '08				Feb '08				Mar '08					Duration Function
		24	31	7	14	21	28	4	11	18	25	3	10	17	24	31	
SUMMARY	Project summary - used for sensitivity calculations															
0010	Design															Triangle(90;100;120)
0020	Test															Triangle(45;50;60)

Figure 10.4 Two activities with duration uncertainty but no correlation

correlated, their variations between projects (represented by the Monte Carlo simulation results) will be unrelated and the correlation coefficient should be zero.

After the simulation the 5000 pairs of durations for the two activities are saved to Excel and the function 'Correl' (short for correlate) is applied to the 5000 pairs of values. The function has calculated the correlation coefficient of -3 percent, slightly negative but effectively zero, approximating complete independence.[4]

PERFECT CORRELATION

Now we want to model how correlation is generated in real projects.[5] We use the risk driver approach to model the way risks actually influence activity durations and cause correlated variables. Assume there is a single risk, 'Technology may differ from our assumptions' as shown in Figure 10.5.

Description	Optimisitic	Most Likely	Pessimisitic	Likelihood
Technology may differ from our assumptions	90.00%	100.00%	120.00%	60.00%

Figure 10.5 One risk driver assigned to design and test activities generates perfect correlation

That risk has a 60 percent probability of occurring and, if it occurs, has an impact that is randomly chosen for each iteration between 90 percent and 120 percent. For any iteration, the duration of each activity to which it is assigned is affected by the risk since it is multiplied by:

- 1.0 if the risk does not occur on that iteration. The durations multiplied by 1.0 are not affected. For 2000 of the 5000 iterations the risk does not occur, in this example.
- A random number chosen between .9 and 1.2 with a most likely value of 1.0 if the risk does occur. For 3000 of the 5000 iterations this number is chosen and applied to all activities to which it is assigned.

Suppose that both Design and Test activities are affected by that risk (we have deleted the three-point estimates for this exercise). Generation of a database of Design and Test activities for 5000 'projects' using the Monte Carlo simulation shows that these two activities that are affected by the same risk are highly correlated. The correlation coefficient calculated using the simulation results data when a single risk and no other affects the two activities is 99 percent, effectively perfect correlation.

---

4   The 5000 pairs of duration values generated by the simulation can be saved as a CSV file for importing to Microsoft Excel®. The resulting durations for the two activities are saved and imported to Microsoft Excel® and the 'correl' function is applied.

5   This example uses Pertmaster from Primavera because of its risk factors capability.

PARTIAL CORRELATION

We have seen how to generate pairs of duration values for two activities that are not correlated and for two activities that are perfectly correlated. What happens if the activities are affected by the same risks and also by other risks? They should be correlated but the presence of other risks that are not common to the two activities will moderate the correlation coefficient. These other risks may be called 'confounding risks.'

Partial correlation is the most usual case found in real projects. Usually two activities are affected by different combinations of risks. Even if there are some risks that are common to them, other risks are not. This is how partial correlation is generated in real projects and we can approximate that effect with the simulation results. In Figure 10.6 we introduce two confounding risks that will be assigned one to Design and one to Test.

	Description	Optimisitic	Most Likely	Pessimisitic	Likelihood
1.	Technology may differ from our assumptions	90.00%	100.00%	120.00%	60.00%
2.	Design Risk A	90.00%	100.00%	120.00%	60.00%
3.	Test Risk B	90.00%	100.00%	120.00%	60.00%

Figure 10.6 Three risks with the same parameters

While these risks have the same parameters they are not themselves correlated. For instance, even though each has a 60 percent probability of occurring, they will not all occur in the same iterations. Also, if the computer were to select a factor of, say, 115 percent for Technology Risk on a specific iteration there is no reason that Design Risk A and/or Test Risk B will also choose the same value or, in fact, even exist on that iteration. We have used the same parameters mainly to keep the example as simple as possible.

Assume the assignment of the three risks to the two activities is displayed in Figure 10.7. The only common risk is the uncertainty in the technology, and the other two risks are those that affect only Design or only Test.

Activities	Risks Assigned		
	Technology Risk	**Design Risk A**	**Test Risk B**
Design	X	X	
Test	X		X

Figure 10.7 Assignment of three risks will generate partial correlation of 50 percent

With these three risks assigned so that only one is common to the two activities, the measured correlation coefficient is calculated at 50 percent. (In Figure 10.3 the calculation with a similar configuration was 48 percent.) From this exercise we see that correlation generated by common risks and moderated by the presence of other risks that are not common to the pair of activities considered is neither perfect (1.0) nor independent (0.0). This is the most usual case in projects.

The Effect of Correlation on Schedule Risk

Why should we care about whether our activities are correlated or not? The short answer is that correlation makes it more possible for very long and, somewhat less likely, very short schedules. We are sensitive to schedules overrunning and we recommend establishing a contingency reserve of time to protect against that eventuality. Correlation increases the amount by which they could overrun and hence increases the time contingency reserve needed. For this reason correlation must be examined and included in our analysis.

Correlation is more likely to make the project schedule very long than very short, although both are possible. The asymmetry is caused because of schedule logic:

- If any path is very long the project schedule will be very long.
- If any path is very short the project schedule may not be short because there may be another path with little total float that will become the critical path. If one path is very short the schedule completion date will be determined by the near-critical path rather than the path with correlated activities, keeping the project from finishing much earlier.

Correlation has more relevance to an organization that 'lives in the tails of its distribution' and chooses a target that is, say, the P-80 rather than one that makes decisions based on average or the median P-50 results. This phenomenon will be shown below. A simple one-path three-activity example will help us to see this impact. The schedule in Figure 10.8 will be used.

ID	Task Name	Rept ID	Min Rdur	ML Rdur	Max Rdur	Qtr 4	Qtr 1	Qtr 2	Qtr 3	Qtr 4	Qtr 1	Qtr 2	Q
0	**Correlated Project**	2	**0 d**	**0 d**	**0 d**								
1	Start	0	0 d	0 d	0 d		1/9						
2	Hardware Design	0	85 d	100 d	130 d	1/9	4/17						
3	Hardware Fabrication	0	255 d	300 d	390 d		4/18			2/11			
4	Hardware Test	0	75 d	100 d	200 d					2/12	5/22		
5	Finish	0	0 d	0 d	0 d						5/22		

Figure 10.8 Simple schedule for testing correlation

Without correlation among Design, Fabrication and Test the schedule risk result is shown in Figure 10.9.[6]

---

6   For this exercise we use Risk+ from Deltek that simulated Microsoft Project® schedules. Risk+ employs Pearson Product Moment correlation, which calculates the correct impact on schedule risk.

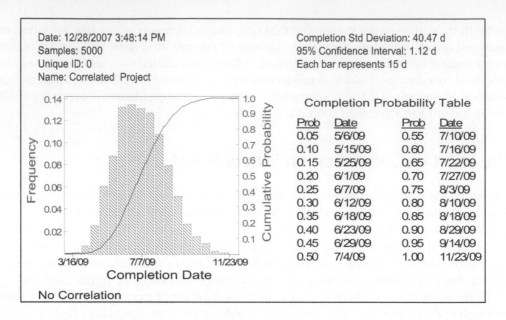

Date: 12/28/2007 3:48:14 PM
Samples: 5000
Unique ID: 0
Name: Correlated Project

Completion Std Deviation: 40.47 d
95% Confidence Interval: 1.12 d
Each bar represents 15 d

Completion Probability Table

Prob	Date	Prob	Date
0.05	5/6/09	0.55	7/10/09
0.10	5/15/09	0.60	7/16/09
0.15	5/25/09	0.65	7/22/09
0.20	6/1/09	0.70	7/27/09
0.25	6/7/09	0.75	8/3/09
0.30	6/12/09	0.80	8/10/09
0.35	6/18/09	0.85	8/18/09
0.40	6/23/09	0.90	8/29/09
0.45	6/29/09	0.95	9/14/09
0.50	7/4/09	1.00	11/23/09

No Correlation

Figure 10.9 Results without correlation

Without correlation:

- the P-80 date is August 10;
- the schedule date of May 22 is about 10–15 percent likely;
- notice the 'Completion Std. Deviation' of 40.47 days. The standard deviation relates to the dispersion or the spread of the distribution measured from the mean date of July 7 (not from its schedule date of May 22).[7] The standard deviation of the distribution without correlation is about 40.5 days. The average or expected completion date, shown at the bottom of the chart, is July 7.

Effect of Correlation on a One-Path Schedule

Usually correlations are displayed in a Correlation Matrix. That matrix displays the correlation coefficients between any pair of activities. Now, suppose that strong correlation exists between Design, Fabrication and Test activity pairs. There are three pairs of activities:

1. Design-Fabrication;
2. Design-Test and;
3. Fabrication-Test.

---

7   If the distribution were a standard Normal or Gaussian distribution (the familiar 'bell curve') about 68 percent of the distribution would be between July 7 plus or minus 40.5 days.

In the example Table 10.1 shows that we have assumed that each pair of activities is highly correlated at 90 percent.

In this simple example the correlation between pairs of activities is equal. This does not need to be the case, since within limits (described below) these correlation coefficients can take on different values if some pairs are more strongly correlated than others. Of course we may also expect that some activities are not correlated at all and that their correlation would show up in the matrix as zero. The simple three-activity one-path schedule is simulated with these high correlation coefficients and generates the results shown in Figure 10.10.

Table 10.1   Correlation matrix with high correlation between each pair of activities

Correlation Matrix			
	Design	**Fabricate**	**Test**
Design	1.0	0.9	0.9
Fabricate	0.9	1.0	0.9
Test	0.9	0.9	1.0

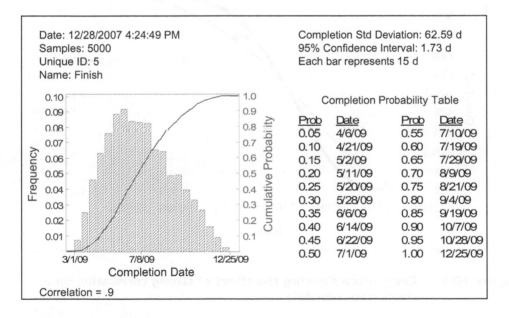

Figure 10.10   Schedule risk with high correlation coefficients

Notice that with high correlation:

- the P-80 date is September 4 compared to August 10 without correlation;

- the probability of finishing on or before the schedule date of May 22 increases to 25 to 30 percent compared to 10 percent–15 percent without correlation. Correlation with a one-path schedule spreads both tails of the result distribution since there are no parallel paths to keep the project from finishing earlier;
- the standard deviation with strong correlation is 62.6 days, which can be compared to 40.5 days without correlation;
- the average completion date is July 7 without correlations and July 8. These are effectively the same day.

These findings show that, with a single-path schedule (or proceeding along any uninterrupted schedule path) correlations increase the spread, both longer and shorter, of the distribution but do not change the mean value. If the organization cares about a high likelihood of achieving a date it might choose a P-80 or some other high percentile, and those are definitely affected by correlation. If it schedules to the mean or even the P-50 correlation is not as important.

The comparison of these two results, without correlation and with strong correlation, is shown in Figure 10.11, with the arrow at the P-80 value that has been our focus.

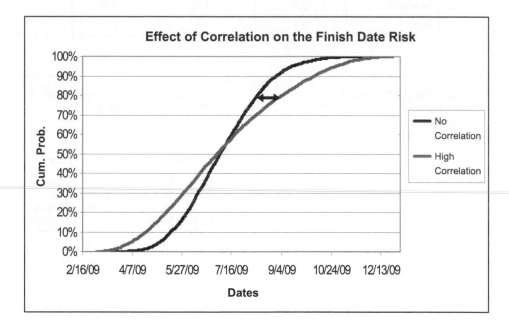

Figure 10.11   Comparison showing the effect of strong correlation on a single-path schedule

The Effect of Correlation on a Two-path Schedule

Most projects do not have a single path, so there is not as much opportunity for earlier completion dates as shown in the single-path project above. Projects with parallel paths find that correlation mainly pushes the distribution toward overruns. We cannot take full advantage of optimistic results on one path because uncorrelated parallel paths may keep the project completion date from being earlier.[8]

Let us introduce a Software path that has activities that are correlated with each other but not with those of the Hardware path. Assume that this Software path is the same in every way as the Hardware path—there are three activities linked finish-to-start with the same durations and risk distributions as on the comparable hardware activities. This new project is shown in Figure 10.12.

To begin the analysis we simulate the schedule without correlation. The results are shown in Figure 10.13.

ID	Task Name	Rept ID	Min Rdur	ML Rdur	Max Rdur	Curve
1	Start	0	0 d	0 d	0 d	0
2	**Hardware**	**0**	**0 d**	**0 d**	**0 d**	**0**
3	Hardware Design	0	85 d	100 d	130 d	2
4	Hardware Fabrication	0	255 d	300 d	390 d	2
5	Hardware Test	0	75 d	100 d	200 d	2
6	**Software**	**0**	**0 d**	**0 d**	**0 d**	**0**
7	Software Design	0	85 d	100 d	130 d	2
8	Software Fabrication	0	255 d	300 d	390 d	2
9	Software Test	0	75 d	100 d	200 d	2
10	Finish	2	0 d	0 d	0 d	0

Figure 10.12   Two-path schedule for correlation comparisons

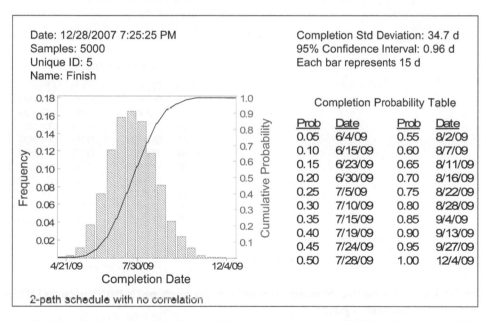

Date: 12/28/2007 7:25:25 PM
Samples: 5000
Unique ID: 5
Name: Finish

Completion Std Deviation: 34.7 d
95% Confidence Interval: 0.96 d
Each bar represents 15 d

Completion Probability Table

Prob	Date	Prob	Date
0.05	6/4/09	0.55	8/2/09
0.10	6/15/09	0.60	8/7/09
0.15	6/23/09	0.65	8/11/09
0.20	6/30/09	0.70	8/16/09
0.25	7/5/09	0.75	8/22/09
0.30	7/10/09	0.80	8/28/09
0.35	7/15/09	0.85	9/4/09
0.40	7/19/09	0.90	9/13/09
0.45	7/24/09	0.95	9/27/09
0.50	7/28/09	1.00	12/4/09

2-path schedule with no correlation

Figure 10.13   Two-path schedule with no correlation exhibits the merge bias

---

8   This is the same phenomenon as the merge bias introduced in Chapter 6, with correlation introduced.

Without correlation the merge bias affects the result as shown in Chapter 6. Notice that with the merge bias:

- The P-80 is August 28, not August 10 as it is for the one-path uncorrelated schedule (and as it is for each of the Hardware and Software paths of this two-path schedule, individually). Compare these results with those in Figure 10.9, above.
- The chance of May 22, which is still the schedule date, is not impossible but is less than 5 percent, not 10 percent–15 percent as in the one-path uncorrelated schedule.
- The mean is July 30 rather than July 7 found for the one-path uncorrelated schedule.
- The completion standard deviation is only 34.7 days, somewhat less than the 40.5 days of the uncorrelated one-path schedule. While the merge bias moves the S-curve to the right it also straightens up that curve showing less range between optimistic and pessimistic results.

To see the impact of correlation on a multi-path schedule we correlate the three Hardware activities 90 percent and the three Software activities 90 percent but we do not correlate any activities between the two paths. The entire schedule does not get the benefit from the correlation of short durations on one path because the other path may be longer and determine the schedule's date.

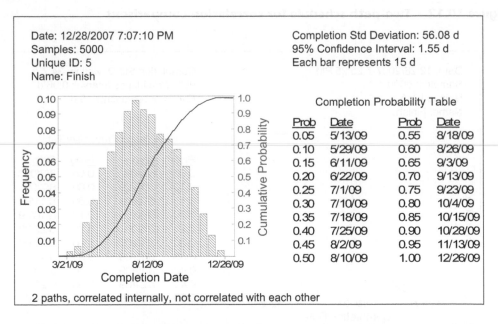

Figure 10.14   Two paths: Strong correlation on each path but no correlation between paths

Notice that, for the two-path correlated schedule, compared to the one-path correlated schedule:

- The P-80 date is October 4 in Figure 10.14 whereas with one path correlated, it was September 4 in Figure 10.10, above. Long projects commonly result from both the merge bias and correlations.
- The standard deviation is now 56.1 days compared to 62.6 days with one path and correlated activities. The merge bias straightens up the S-curves.
- The chance of the schedule date of May 22 is between 5 percent and 10 percent. With one path and correlation the schedule exhibited more than 25 percent probability of the schedule date. The benefits of correlation on the early side are much less if there are multiple paths than if there is just one path because each path may delay the project by itself.
- The average date is August 12 for the correlated schedule compared to July 8 for the correlated one-path schedule. In contrast to the one-path result, where there is more than one path the mean is affected by correlation.

The effect of correlation on a two-path schedule is more dramatic than it is on a one-path schedule because of the merge bias. This is shown in Figure 10.15 in comparison with Figure 10.11 above.

Notice that for schedules with multiple paths the effect of correlation in producing long schedules is more pronounced (mean and P-80) than it is for producing short schedules with multiple paths. The merge bias keeps the correlation from producing short schedules although it permits longer schedules. This result gives us both:

- a pessimistic outlook with correlated schedules;
- a good reason to be alert to identify, calibrate and include correlation in the simulation when we see it in multi-path (that means, all real) schedules. Otherwise we will underestimate schedule risk at the higher probabilities where prudent organizations focus.

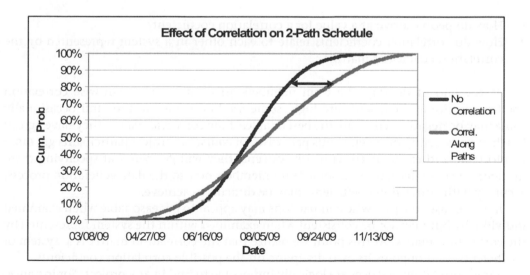

Figure 10.15   Impact of correlation on two-path schedule

Specifying Correlation Coefficients—The Inconsistent Correlation Matrix

In traditional schedule risk analysis, the values of correlation coefficients are usually determined in interviews or risk workshops where project participants talk about risks and their impacts on the project schedule. It is rare that the participants in these meetings have historical data that can be used to compute the correlation coefficients. And yet, not having data does not mean that correlation is zero. The results above show that we should not ignore correlation. Even if they have no historical data the project participants need to become adept and comfortable estimating correlation coefficients using their expert judgment.

As mentioned above, the teams look for:

- common risks driving the durations of pairs of activities;
- risks that are important in determining activity durations;
- risks that show significant variability.

There are some difficulties in developing correlation coefficients using expert judgment, however:

- the team members are often not familiar with the concept of correlation;
- the calibration of correlation coefficients is difficult. For instance, what coefficient should be assigned when someone believes there is 'strong' correlation between a pair of activities?
- two people might agree that correlation is 'strong' but give a different coefficient?
- people often argue about whether correlation is 'strong' or 'moderate' but have the same coefficient in mind?

The main questions are:

- How do people arrive at a value for a correlation coefficient?
- How do correlation coefficients relate to each other in a system represented by the correlation coefficient matrix?

At the end of the day, the team members arrive at a set of pair-wise correlation coefficients. Some coefficients are high, some are low, some are zero and occasionally some may be negative. These are the best or at least the consensus values they have agreed to when considering the coefficients pair-wise. Of course, as project participants get more comfortable and experienced with risk concepts they will get better at specifying these numbers. At the outset, or with some team members new to the data generation process, accuracy with correlation coefficients may be difficult to achieve.

In some cases the pair-wise comparisons may appear to be reasonable when examined individually but may not be consistent when examined within the system represented by the correlation matrix. At this point we realize that the correlation matrix is a 'system of coefficients' that imposes its own discipline on the possible correlation coefficients.

Some correlation matrices are logically impossible to find in any project. For instance, even though project participants may settle on the correlation coefficient matrix shown in Table 10.2, it is impossible to imagine 100 projects, each one on a line in a spreadsheet with durations of Activities A, B and C arrayed in columns where:

- Activity A is strongly correlated with Activity B 90 percent.
- Activity A is strongly correlated with Activity C 90 percent.
- Activity B and Activity C are only correlated 20 percent.

Table 10.2 An inconsistent correlation matrix

Correlation Matrix			
	Design	**Fabricate**	**Test**
Design	1.0	0.9	0.9
Fabricate	0.9	1.0	0.2
Test	0.9	0.2	1.0

The team may agree on these correlation coefficients when they are discussed one at a time, but the coefficients have to be feasible as a system.

The problem can be seen immediately once the coefficients are arrayed in the matrix. If Design and Fabricate are strongly related, and Design and Test are strongly related, then Fabricate and Test must be at least correlated with moderate strength. The matrix is impossible—the coefficients are inconsistent within the system. There are three ways this coefficient matrix is impossible:

- the correlation of 20 percent between Fabricate and Test is too low;
- the coefficients of 90 percent between the other two pairs are too high; or
- both the high coefficients are too high and the low coefficient is too low.

In any case, this correlation matrix is not internally consistent. This set of correlation coefficients will not happen on real projects. Yet the project team might have settled on these values when discussing pair-wise correlations. The system shows a problem that can occur with coefficients that are determined by expert judgment.

Happily there is a test for the correlation matrix internal inconsistency. The Eigenvalue Test is based on the characteristics that such a matrix (square, values of 1.0 down the principal or main diagonal and mirror off-diagonal coefficients) must have. The Eigenvalue Test shows that this matrix is a mathematical impossibility and hence is a real impossibility as well.

Simulation software that offers proper correlation also offers the Eigenvalue Test. It performs this test before it begins the simulation and alerts the user if the correlation coefficient matrix is inconsistent. Most software packages will offer to fix the correlation matrix by creating a new matrix that is as close as possible to the one specified but which just barely passes the Eigenvalue test. Some of these software packages will indicate to the user which fixes were made so the user can review them and determine if those changed coefficients are acceptable.

An example of providing the revised correlation matrix is Crystal Ball®[9], a Microsoft Excel® simulation package often used for financial or cost risk analysis. This software is used for the purposes of this experiment because it tells the user what the resulting 'near-by' matrix is that passes the Eigenvalue Test. Putting in the matrix shown above returns the message shown in Figure 10.16 from Crystal Ball®. The program will not perform the simulation until the matrix is fixed, either by an internal algorithm or by the user, to pass the Eigenvalue test.

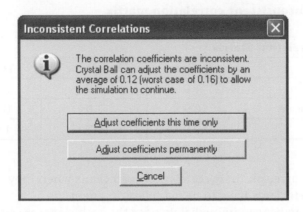

Figure 10.16   Message from failing the Eigenvalue Test (Crystal Ball®)

Choosing the 'Adjust coefficients permanently' option allows the user to see the revised matrix. In this case the coefficients have been changed by that software as follows:

Table 10.3   Correcting an inconsistent correlation matrix (by Crystal Ball®)[10]

Correlation Matrix			
	Design	**Fabricate**	**Test**
Design	1.0	0.9	0.9
Fabricate	0.9	1.0	0.2
Test	0.9	0.2	1.0
Has been changed by Crystal Ball® to:			
	Design	**Fabricate**	**Test**
Design	1.00	0.74	0.74
Fabricate	0.74	1.00	0.24
Test	0.74	0.24	1.00

9   Crystal Ball® is made and sold by Decisioneering, a unit of Oracle.

10   Different software uses different algorithms to find a nearby correlation matrix that passes the Eigenvalue Test. Some do not tell you what they did. Risk+ performs the Eigenvalue test and offers to revise the correlation matrix to pass that test. The revised correlation matrix values can be printed out to a .csv file.

Notice that the high correlation values of .90 have been reduced to .74 and the low correlation value of .20 has been increased to .24. This matrix just barely passes the Eigenvalue Test, but it may not be agreeable to the project team. Different software packages will implement their own algorithms that may result in a different corrected matrix. Using the revised matrix we can proceed with the simulation.

Risk+ will also adjust the correlation coefficient matrix to pass the Eigenvalue Test. The results are somewhat different from those of Crystal Ball®.

Table 10.4   Correcting an inconsistent correlation matrix (By Risk+)

Correlation Matrix			
	Design	**Fabricate**	**Test**
Design	1.0	0.9	0.9
Fabricate	0.9	1.0	0.2
Test	0.9	0.2	1.0
Has been changed by Risk+ to:			
	Design	**Fabricate**	**Test**
Design	1.0	0.84	0.84
Fabricate	0.84	1.0	0.27
Test	0.84	0.27	1.0

Passing the Eigenvalue Test does not ensure that the correlation coefficients are the best ones to use or that they are accurate. It just ensures that they are not mathematically impossible. This is a low threshold, but one that we should not trip over. When the software tells the user that the pair-wise correlation coefficients when put into the correlation coefficient matrix fail the Eigenvalue Test, it is best to identify the inconsistency in the correlation matrix and then go back to the project team to re-estimate the coefficients with system integrity in mind. The team members may have a different and preferable solution to the one that the computer calculates.

Pearson Product Moment versus Spearman Rank Order Correlation Approaches

There are two concepts of correlation that are in use with retail simulation software, the Pearson Product Moment approach and the Spearman Rank Order approach. The Pearson Product Moment approach is preferred. It preserves the linear relationship between variables when creating the correlated series. This means that it recognizes that 3 is 1.5 times larger than 2, not only that 3 > 2. This is the way we intuitively think about correlation. Using the Pearson approach the results of the simulation properly accounts for the amount of spread in the inputs and gives more accurate results than the Spearman method. Combining inputs using this approach will result in the proper degree of risk without the underestimate bias present in Spearman.

Some of the retail software that provides Monte Carlo simulation capabilities uses the Spearman Rank Order approach to correlation, an approach that is theoretically inferior to the Pearson Product Moment approach. The Spearman approach may somewhat underestimate the effect of correlation on the result probability distribution—that is, the resulting distribution is narrower than it should be. This is because the Spearman correlation only assures that the values are rank ordered. It preserves the monotonic order of values but does not preserve the linear relationship, the degree to which the outliers are away from the mean. That means that it recognizes that 3 is larger than 2, but it does not recognize that 3 is 1.5 times 2.

Risk+ from Deltek, a software product that simulates Microsoft Project® schedules implements Pearson Product Moment correlation. The algorithm (Price 2002) is a sophisticated application including the Lurie-Goldberg transformation (Lurie and Goldberg 1998) that allows Pearson Product Moment approach to be used with distributions that are not Normal or Gaussian.

Other software packages implement the Spearman approach or some other approach that relates activity durations in a proprietary non-correlation approach. In practice the results from using the different approaches do not seem to differ significantly. However, we should use a calculation method, Pearson, if it is available, because its results would be correct if the data were perfectly accurate.

Summary of Correlation in Project Schedule Risk Analysis

It is a fact of project life that risk is not always random. Sometimes risks can affect more than one activity of a schedule, and cause those affected activities' durations to move together. That means that on any project the activities durations will be long or short within their probability distributions together. If they are long together there is a greater chance for serious overruns because long durations that occur on the same project reinforce each other, making the project very late.

Correlation is imposed on the simulation so that each iteration will represent an internally consistent scenario. If it is improbable that one activity is long while another is short (within their own risk ranges) because they are both influenced by the same risks, then they are correlated. If the duration inputs to the simulations were selected purely at random the relationships between durations would not be enforced and some iterations would be internally inconsistent. Monte Carlo simulation software allows the user to specify correlation so that the inputs to the iterations will be long or short together as observed in real life or at least believed to be so by the project team.

Correlation is a way to evaluate the way pairs of activities' durations will be high or low together on the same project, representing positive correlation that can be strong or weak. (Of course we might have negative correlation, although that is rare in project management.) We used Monte Carlo simulation with uncorrelated activities, perfectly correlated activities and partially correlated activities to synthesize data for 5000 projects. When we computed the correlation coefficient using a spreadsheet function we found coefficients that were close to zero for the uncorrelated example, close to 1.0 for the perfectly correlated example and close to 50 percent for the specific example that had one risk that was common to two activities but two risks that were not common to the pair of activities. Correlation occurs in different strengths and the project team will have to be alert to degrees of correlation.

Unfortunately most organizations do not have historical data from a number of projects from which they can compute correlation coefficients so they must estimate those coefficients using project teams' expert judgment. We saw that this expert judgment can cause inconsistent matrices of correlation coefficients. We introduced the Eigenvalue Test to help the user avoid inconsistent correlation matrices. If the correlation matrix fails this test, the simulation software usually offers an algorithm to find a near-by matrix that passes the test. Risk managers are encouraged to review these adjusted coefficients but to go back to the project team for confirmation and correction.

Using the risk driver method of schedule risk analysis (see Chapter 8) the risks are the basic unit of measurement and correlations arise naturally as the risks are assigned to activities. This approach avoids both the need to specify the size of the correlation coefficients and the possibility that the specified coefficients might be internally inconsistent.

We saw that the impact of correlation on project schedule risk is to spread out the distributions along a path. In the rare single-path schedules or along a single path it will not change the expected value very much and will spread the distributions similarly in the optimistic and pessimist directions. In the more common multiple path projects the impact on the pessimistic or high end of the distribution is more pronounced than for the low or optimistic end because of both correlation and the structure of the schedule—the merge bias is at work. In each of these cases, the P-80 or other percentile results in the upper tail of the distribution will be later and the time contingency reserve needed to provide the client with the required level of confidence will be greater than if correlations are not recognized or not specified.

Finally we mentioned that we were using the Pearson Product Moment approach to correlation and indicated that it is preferable to the Spearman Rank Order approach. Pearson is a linear approach preserving both the rank order and the degree of difference from the mean, which is the preferred approach. Spearman preserves the rank order of values but not the linear relationship between variables. Hence, Spearman theoretically underestimates the impact of correlation on the total schedule risk—the probability distribution of total project completion dates is too narrow under Spearman. Fortunately there is at least one product that has implemented the Pearson Product Moment approach. In practice the comparison of results shows that the difference between approaches may not be substantial.

Correlation is a fact of project life and the project manager who ignores its importance will underestimate the risk to the project schedule.

References

Lurie, P. and Goldberg, M. (1998). 'An Approximate Method for Sampling Correlated Random Variables from Partially-Specified Distributions.' *Management Science* 44(2): 203–218.

Price, J. (2002). An Implementation of the Lurie-Goldberg Algorithm in *Schedule Risk Analysis Space Systems Cost Analysis Group*. Long Beach, CA.

11 Risk Management in the Organization: Identifying the Mature Risk Management Culture

Introduction

There are three main components to successful project schedule risk analysis:

1. We must have a project schedule that adequately represents the project plan. It must be dynamically stable in that it automatically produces the correct dates and critical paths when durations change.
2. The data about project risks that affect the schedule must be as accurate as can be known at the time of its collection. This requires attention to the data collection process and people's ability to describe risks.
3. We need to know and master the tools and techniques of Monte Carlo simulation, using modern software that is available.

Not all Organizations Value Risk Analysis

The overriding success factor for project schedule risk analysis, however, is an organizational culture that is supportive of and values the project risk analysis process.

Organizational risk management maturity starts with the organization's management. Even organizations in which the conduct of quantitative project risk analysis is an established process may only give lip service to schedule risk. Often these same organizations end up arguing that the results (for example, a 10-month delay at the (P-80) must be flawed) rather than trying to learn from them.

Risk management maturity is a cultural attribute that begins at the top. Project risk awareness may not come naturally to the project-oriented organization. A professional schedule risk analysis can be wasted if the organization's culture does not value its contribution or in fact ignores or discounts its message while making decisions about the project. If management is antagonistic towards quantitative risk analysis the project teams will read the message behind their behavior and fully understand that the exercise is opposed. This cultural attitude will destroy the ability to conduct risk analysis of schedules or to benefit from it.

If the organizational risk management culture opposes or cannot see the use of a risk analysis, no application of sophisticated methods, tools or training will be successful. An organization's culture is more important to the success of project risk analysis than are tools, methods and training combined.

Elements of a Risk-Mature Culture

A risk-mature culture will have several, easily observable, elements. If the organization has these attributes, schedule risk analysis can provide a positive contribution to the success of the project and a rewarding experience for all. However, absence of these elements or the presence of a risk-immature culture in the organization may both sabotage the risk analysis exercise and make its practice impossible.

A risk-mature management recognizes that it needs to know about the risks to their project schedules. It is surprising to find out that management, in some companies or government agencies, does not always want to know about the risk to their projects' schedules. For instance, if discovering and acknowledging the risk will put the project in jeopardy, management is often antagonistic toward the concept and conduct of project schedule risk analysis. If a published or announced deadline for delivery may be in jeopardy, some organizations do not want to examine schedule risk too carefully.

Management needs to believe that best-practice schedule risk analysis is a valuable tool of advanced and competent project management. Risk analysis is taught in universities. Practitioners and software developers are advancing the state-of-the-art continuously. We sometimes find out, however, that managers remain skeptical toward or actually oppose risk analysis since 'they have survived for many years without practicing it.'

Management should be open to new ideas that can help manage projects. Some very mature high-level managers have said that: 'We must do this, we have never been able to quantify these risks before and we cannot advance unless we embrace this new methodology.' It is not unusual, however, to find management reacting to unfamiliar techniques such as Monte Carlo simulation as threatening, somewhat mysterious and therefore suspicious. Top managers who are otherwise sophisticated, some who actually are 'rocket scientists,' often become uneasy toward applying statistical concepts, even if they have a background in this approach. This is a barrier to their willingness to accept it.

Management needs to be willing to commit the time and resources to support the risk analysis effort. On some complex projects the conduct of a schedule risk analysis can take as much as 6 weeks. The process includes assessing and fixing the schedule, collecting data from many knowledgeable people, modeling the schedule risk, examining schedule risk mitigation and reporting on the results. The risk analysis process is based in project documents such as the schedule and risk register, that must exist and be consistent with best practices. The process takes hours of staff time in interviews and to develop meaningful risk mitigation actions. The risk analysis may involve hiring outside consultants and acquiring training and computer software. Obviously if there is some resistance to the very concept of schedule risk analysis, the need for this commitment of time and resources may be used as a reason to avoid the exercise altogether.

Management should establish the project risk management function in the organization chart to be independent of the project managers themselves. Of course the project risk analysis cannot be conducted without support of the project managers and their teams; data and elemental project documents such as the schedule must come from the project team. However, if the project manager can direct changes to the analysis in order to support some preconceived result, then the analysis will be compromised and rendered useless. The risk analyst should be set up to report to a senior officer or to a PMO.

Powerful Use of the Schedule Risk Analysis Results

Risk-aware project management should understand how risk analysis results can be used. The schedule risk analysis is conducted on a specific project plan. The analysis results show management where the risks are so that the plan can be improved for better results.

Mature managers will use the risk analysis results as the start of a thorough search for risk mitigations. More mitigation options are available early, in the project planning phase, than later in the project lifecycle, and they have a better chance of success if they are initiated early. For this reason, risk analysis should first be done in the early stages of project planning. Quantitative risk analysis needs at least a summary schedule, but in mature project management organizations these documents appear during or soon after initial conceptual design.

Risk analysis needs to be conducted to form a basis for risk-aware project decisions such as whether to authorize detailed engineering efforts or to sign major contracts for construction, fabrication or procurement. The conduct of schedule risk analysis should be programmed into the project plan including into the schedule and the budget at major decision points.

Even after execution has begun and the plan is set, risk analysis can help manage expectations. If the delivery of the final product is not going to meet the schedule because of risks that now cannot be mitigated effectively, a realistic expectation that may differ from the current (statused) project schedule should result from the schedule risk analysis.

Management needs to distinguish between analysis results and forecasts. The schedule risk analysis results apply to a specific version of the project plan. The analysis results should not come to pass since the project team should use them as the basis for beneficial changes that mitigate the most important risks. The analysis results will only become forecasts if management does nothing and as a result the current plan stands unimproved.

Prerequisites of Schedule Risk Analysis Require Serious Attention

THE PROJECT SCHEDULE

The first essential starting place for a schedule risk analysis is the project schedule itself. An earlier chapter presented scheduling techniques and warned of scheduling weaknesses and abuses. In practice many schedulers are not strong in the scheduling discipline. They seem to view their jobs as making sure some predetermined dates are reflected in the schedule, whether the schedule logic and durations support those dates or do not. Hence the schedule risk analysis always starts with a thorough review of the project schedule. Often in an engagement, errors or lapses are found in the schedule long after the team has 'blessed it,' since the teams rarely understand the project scheduling discipline at a detailed level themselves.

To support project management itself it seems prudent to establish a strong scheduling discipline with sufficient number of trained, experienced and professional schedulers to do the job. In fact it is difficult to see how a schedule risk analysis can be conducted by someone who is not adept at project scheduling.

RISK ANALYSIS DATA

The next building block is the risk analysis data. Risk data collection is not a rote exercise with a set list of questions that must be asked, answered and entered mindlessly into a database. Risk data collection is not a mail-out and mail-back form to be filled out and typed in. Rather risk data can be collected only with in-depth interviews by experienced interviewers of project participants and others who are knowledgeable about and have experience in the subject matter of the project.

The selection of the interviewees is important to the success of risk data collection. There should be a good mix of high-level interviewees and discipline leads involved in the project. The experience of the interviewer in projects and in project risk is also important; risk data gathering is more of an interpersonal exercise than a technical one. Gaining the trust of the interviewees is a key to success. Also, experience in data gathering will help the interviewer-analyst distinguish between data that will be useful and data that, for some reason such as interviewee bias, will not be useful. The ability to gather good data also implies that on occasion some data and interviewees must be ignored and other sources of data must be sought.

MASTERY IN THE CONCEPTS AND TOOLS OF SCHEDULE RISK ANALYSIS

The final building block is the mastery of the concepts and tools of the schedule risk analysis techniques, the subject of several chapters. These techniques are usually built around sophisticated software. Training in specific software packages, specifically commercially-available Monte Carlo simulation programs, is important. Because software and methods change, sometimes at a rapid pace, building liaisons with software manufacturers is often useful. The software companies usually appreciate hearing from constant and serious users when they want to determine their development strategy. Attending and presenting at software user conferences can be important as well. Because these tools are often specific to particular scheduling packages it can be useful to be familiar with and own several different simulation packages.

The Risk Culture is the Key Success Factor

The whole schedule risk analysis enterprise rises or falls on the type of culture within which it is carried out.

It is not easy to convert the minds and hearts of senior managers to a new method that may put their projects in jeopardy. It is sometimes difficult to get traditional project managers to use risk analysis results to improve their projects when they do not see realistic expectations as serving their own interests. Once a plan has been adopted it is sometimes difficult to get project managers to accept the need to plan for risk mitigation, even if that may improve project performance. Often promises are made based upon considerations beyond the project that will deliver on those promises. It is in this case that risk analysis is most needed but sometimes most opposed by management.

Organizations cannot be driven to adopt a risk mature culture; they must see the benefit and want to allocate time and resources to risk analysis. One possible approach is to identify a champion of risk analysis and to introduce the practice of risk analysis through that person on a specific project. You may find, particularly in some of the more hierarchical cultures, that the support and endorsement of these methods by a senior leader can soon bring the organization around to the practice of schedule risk analysis.

In some cases agents outside of the organization can mandate introduction of these methods. In recent years government oversight agencies have made a point of requiring risk analysis on some categories of projects. Still, the widespread acceptance of these methods by the agencies and contractors that are subject to this oversight is not yet secure.

It is a joy to be part of a risk analysis in a risk-mature organization that embraces the exercise and uses it in the way it was intended; that is to help focus on risk mitigation that may be available early in the project life cycle. Too often a risk analysis is requested once a project is already in trouble, but even in that situation it can provide benefit in developing realistic expectations. Risk-immature organizations either sabotage the efforts or complain that the results cannot be accurate and useful. The risk-mature culture finds ways to benefit from schedule risk analysis and to use it to move forward toward better project results.

1 *The Problem with PERT*

PERT stands for the method of analyzing project schedules called 'Program Evaluation and Review Technique.'[1] According to the US PERT Orientation and Training Center (POTC) PERT was developed in the late 1950s to account for the impact of project schedule risk on the planned completion of major projects. (USPOTC 1962) Unfortunately, PERT always underestimated schedule risk. The problem is that PERT has always ignored important risk at the merge points contributed by the shorter path. This is the merge bias described in Chapter 6. PERT instead focused on the one longest or PERT-critical path (Hulett 2007).

Because PERT ignores the merge bias, it underestimates schedule risk at each merge point, systematically ignoring the possible risk that may be caused by PERT-slack paths. For this reason PERT results are not just inaccurate—they are inaccurate in a predictable direction but by an unpredictable amount. Many in the project management profession do not understand this problem, although it has been around and discussed in popular scheduling books since the mid-1960s.

In most schedules it is predictable that the schedule risk is greater than the PERT results would indicate. The conditions under which PERT is accurate (single-path schedules or slack paths with more float than risk at every merge point and organizations willing to adopt early starts strategies) are so limited that it can be said that PERT is almost always wrong, that it almost always underestimates schedule risk. Decisions based on PERT results are potentially wrong because they do not appreciate the full extent of the schedule risk. Monte Carlo on the other hand is an international best practice and handles the merge bias correctly.

This section describes PERT and illustrates its lack of recognition of the merge bias. It shows that Monte Carlo simulation produces the correct answers without this flaw. Monte Carlo simulation is the modern way to evaluate project schedule risk. PERT, which was a significant advance when it was first applied and up until the Monte Carlo alternative was made widely available, has been discredited among those who practice schedule risk analysis.

The Project Evaluation and Review Technique or PERT— Background

The Program Evaluation and Review Technique, commonly abbreviated PERT, was a model for project management invented by Booz Allen Hamilton, Inc. under contract to the United States Department of Defense's Navy Special Projects Office in 1958 as part of the Polaris mobile submarine-launched ballistic missile project. This project was a

---

1   PERT has been applied to cost uncertainty as well, so there is also PERT Cost.

direct response to the Sputnik crisis. DuPont Corporation's Critical Path Method (CPM) of scheduling was invented at roughly the same time as PERT (Wikipedia 2007).

There were certain fundamentally appealing features of PERT:

- PERT went beyond the standard and still-used today deterministic scheduling approaches in order to include and estimate the impact of risk in the schedule.
- PERT is intuitive and easy to understand, even for a non-statistician.
- PERT used simple computational methods available to most people. (Of course before the easy access to computers, even PERT calculations could be laborious.) Hence, PERT was commonly used in the DOD and other government agencies. (DOD and NASA 1962; USPOTC 1962).
- PERT used a statistical method more properly known as the Method of Moments (MOM). The MOM is consistent with statistical principles and is useful for computing the probabilities down a column of numbers or, equivalently, along a single path of activities linked by finish-to-start logic. Hence, the MOM has been used extensively in computing cost risk where the cost model is often simply adding up a column of numbers. The MOM provides accurate answers on schedules only when the project has a single path or where PERT slack paths have more float than risk.

The PERT developers probably reasoned that: (1) there may be one dominant schedule path and the others would not have enough risk to matter, and (2) the MOM was a simple tool that made risk analysis available to all. They were right that PERT was better than relying simply on the accuracy of uncertain deterministic schedule durations. They did not realize that PERT had problems with a real schedule that includes parallel paths and merge points, however.

The PERT method of approaching schedule risk has its adherents even today. The fact that PERT systematically underestimates project schedule risk when it is used as it was intended has not penetrated to all practitioners, textbooks or journals. Even some people who admit PERT's inaccuracy for project schedule risk analysis claim it has use if you 'just want to get a feeling for the results.' This attitude is perhaps a reaction to a perceived mystery about Monte Carlo simulation techniques and/or a reluctance to buy and learn new software (PERT can be implemented in a spreadsheet), The fact that the merge bias is a little sophisticated has also caused some practitioners to ignore this crucial issue.

The appeal of PERT cannot be because it is easy to collect the data used by this older method. PERT uses the same three-point estimates that are used in the traditional Monte Carlo simulations and are based on activity duration ranges to generate the probability distribution for the schedule activities. Since gathering the risk data is by far the most labor-intensive part of the schedule risk analysis endeavor, using PERT actually saves almost no time or resources. Monte Carlo simulation, which actually works to compute risk correctly at the merge point, is the preferred solution over PERT, which is not accurate. It is puzzling why, after spending the time to collect the data, some people would apply a failed approach. The best conclusion is that the adherents of PERT do not know its problems, since it is difficult to imagine someone who knows the problems would specifically decide to use PERT anyway.

Concerns About the Accuracy of the PERT Methodology Arose Early

Soon after the development of PERT, in the early 1960s, two researchers from the RAND Corporation in Santa Monica, CA identified a problem with PERT's calculations at the merge points of schedules. (MacCrimmon and Ryavek 1962). That problem was associated with PERT's selection of one PERT-critical path as the only path to analyze. The PERT-critical path is the path that is the longest through the schedule when the activity durations are recalculated using information about their optimistic and pessimistic values.

The problem that MacCrimmon and Ryavek detected involves the possibility at merge points that, under some circumstances, a PERT-slack path maybe longer than the PERT-critical path. If there are other paths contributing to the risk of the schedule, they realized, ignoring the contribution of those paths to project risk was a mistake. This concern is exactly the same as the merge bias, which has been described in Chapter 6.

One of the early adopters of and leaders in network-based scheduling was Russ Archibald (Archibald and Villoria 1967). Archibald and Villoria wrote a path-breaking book that introduced Precedence Diagramming Method (PDM) to many project managers. In an appendix to that book, the work by MacCrimmon and Ryavek was reprinted and made accessible to many people.

We now have sophisticated Monte Carlo simulation software designed to simulate the risk in a schedule as well as general access to fast computers needed to run the simulations. Monte Carlo simulations address and compute the merge bias correctly whereas PERT did not. PERT has fallen into disuse among risk management experts. Unfortunately, not everybody got the memorandum about PERT's shortcomings, and PERT still crops up in practice and the professional literature. Fortunately, the Monte Carlo simulation technique provides a way to examine risks that avoids the problems of PERT.

The Basis of PERT

There is a basic misconception about PERT in the project management community. That misconception arises from the use of the term 'PERT' in ways not originally intended by the developers. Some people confuse PERT with the application of three-point estimates to cost or schedule durations to improve the accuracy of the deterministic estimates. This confusion is reinforced by common textbooks and even by some software companies: Microsoft(R) Project has a 'PERT Tool' that computes three-point estimates, essentially implementing only part of PERT.[2] It is because of the common confusion of PERT and three-point estimating of deterministic durations and costs that we need to explain the true PERT.

PERT used a now-standard three-point estimate of activity duration to represent our lack of certainty in the duration estimates (Wikipedia 2007):

---

2   The use of three-point estimates to improve the accuracy of deterministic estimates of durations or of costs is encouraged. This method is included in the Time management chapter of the PMBOK® Guide of PMI, for instance. PMI (2004). *A Guide to the Project Management Body of Knowledge*. Newtown Square, PA, Project Management Institute.

- Optimistic time (O): the minimum possible time required to accomplish a task, assuming everything proceeds better than is normally expected.
- Pessimistic time (P): the maximum possible time required to accomplish a task, assuming everything goes wrong (but usually excluding major catastrophes).
- Most likely time (M): the estimate of the time required to accomplish a task, assuming everything proceeds as normal. This is the duration more likely than any other to occur.
- Expected time (TE): the average or mean estimate of the time required to accomplish a task, assuming everything proceeds as normal (the implication being that the expected time is the average time the task would require if the task were repeated on a number of occasions.

Typically PERT assumed the beta distribution for the activity durations. For the beta, an approximation of TE = (O + 4M + P) / 6. (The form of the probability distribution is not important to an understanding of PERT.) A triangular distribution will be used, for the reasons stated below, so the formulae used will be appropriate to that distribution.)

- PERT critical path: the longest possible continuous pathway taken from the initial event to the terminal event using Expected time, TE. The completion date of this path was assumed to be the average project finish date, considering schedule risk.
- PERT also calculated the standard deviation ('sigma') of each activity along the PERT critical path, assuming a beta distribution and approximating the value as Sigma = (P-O) / 6. (As with TE, if we use the triangular distribution we will need to use a different formula, shown below, for the standard deviation that is correct for that distribution.)
- A well-understood MOM formula for combining activity variances (the square of their standard deviations) along a single path allows estimation of the standard deviation of the total project completion date.
  These steps are illustrated in a table presented below.

A DETAIL: USE BETA OR TRIANGULAR DISTRIBUTION TO COMPARE PERT AND MONTE CARLO?

The PERT assumption of the beta distribution is an unimportant detail in the critique of PERT. In this discussion we will assume the triangular distribution so we can compare, as apples to apples, the results from PERT and Monte Carlo. Why do we assume a triangular distribution when PERT recommended a beta? The reason is three-fold:

1. 'The beta distribution' does not exist. There are many beta distributions that can be consistent with any three-point estimate. A beta can be tall and skinny (high kurtosis)[3] or short and fat (low kurtosis) depending on the values of the shape parameters. Two such beta distributions are shown in Figure A.1. The typical PERT user did not know of this problem and assumed that the formulae were accurate for 'the beta distribution.'

---

3   In *probability theory* and *statistics*, kurtosis is a measure of the 'peakedness' of the *probability distribution* of a *real-valued random variable*. Higher kurtosis means more of the *variance* is due to infrequent extreme deviations, as opposed to frequent modestly-sized deviations. Wikipedia (2007). Kurtosis.

Today that distribution is sometimes called the 'BetaPERT' distribution. The software manufacturers generally choose the shape parameters and hence determine the shape of the beta distribution for the user.

2. The equations proposed by PERT for the mean and standard deviation of 'the' (in quotes) beta distribution are easy and convenient but they are not exact or even accurate.

3. The triangular distribution, by contrast, is fully specified by the three points, O, M and P. The equations for the mean and the standard deviation of the triangular distribution for any triangular are exact and well-known.

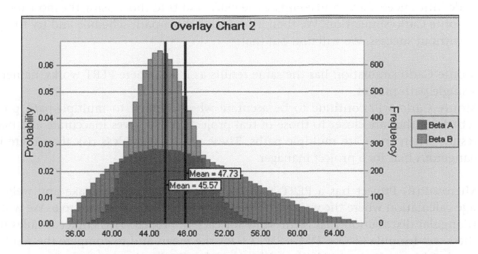

Figure A.1   Comparison of Two Beta Distributions with the same O, M and P values

All Monte Carlo simulation programs have the triangular as an option, and we know it is the same triangular that we assume for PERT in this section. For this reason we can be sure that for the comparison between PERT and Monte Carlo results, we are using the same assumptions. Hence, we use the triangular distribution so that the differences noted in the comparisons of results reflect the methods, not possible inconsistencies the distributions.

The equation for the triangular distribution mean is:

$$\text{Triangular TE} = (O + M + P) / 3.$$

The equation for the triangular distribution standard deviation is a little more complex but still manageable:

$$\text{Triangular Variance} = ((P\text{-}M)^2 + (P\text{-}M) * (M\text{-}L) + (M\text{-}L)^2) / 18$$
Where: Variance is the square of the Standard Deviation

Having calculated the mean and the variance of the PERT critical path, all that is needed is a couple of formulae deriving the mean and standard deviation of a path of activities from the mean and standard deviation of the activities' durations and a table of the normal or some other assumed distribution shape to calculate the probability of the schedule finishing on any date or earlier.

Illustration of PERT When it Works: The One-Path Schedule

From the discussion above it should be clear that PERT's problems arise when there are multiple paths and merge points at which more than one path can contribute to the project risk. To illustrate PERT when it does work, we will use the simple single-path schedule introduced above and compare the PERT results to those using the more modern tool, Monte Carlo simulation. We then introduce a three-path schedule and try to apply PERT, without success. We will find out that:

- Monte Carlo simulation has the same results as PERT where PERT works, namely in a single path project.
- Monte Carlo will continue to be accurate when applied to multiple-path project schedules that are closer to those of real projects. PERT gives inaccurate and biased results when there are multiple paths. These results show less risk than there is, a dangerous bias for a project manager.

Microsoft(R) Project has a PERT Tool that allows the user to make any weighted average calculation where the weights on the three-point estimates sum to 6. We will use the triangular distribution that is the same for each approach. This choice enables us to use the exact formula (2*O + 2*M + 2*P) / 6 for the mean and to compare the results to those using Monte Carlo simulation with triangular distributions.

Assume a single-path schedule introduced in Chapter 6 and shown in Figure 6.2. It has four activities and their durations:

1. Design 50 days
2. Build 100 days
3. Test 35 days
4. Ship 15 days.

This project is nominally estimated to take 200 days as shown in Figure 6.2 of Chapter 6.

Now, assume the three-point risk ranges shown in Table A.1 and the triangular distribution. Applying the PERT Tool in Microsoft Project®, the results using PERT appear in Table A.1

Notice that PERT (using triangular distributions) calculates the duration of this project to be 220 days, not 200. The finish date implied by PERT is January 6 as shown in Table A.1, not December 17 as shown in Figure 6.2 in Chapter 6. The average risk in the activity durations along the single path is taken into account in the average completion date.

This calculation is as far as the PERT tool takes us. However, PERT went further to help us understand the probability distribution of the schedule. We cannot determine how likely the schedule date or any other date other than this average date of January 6 is without estimating the shape and parameters of the probability distribution. PERT

Table A.1   PERT approach in Microsoft Project® using triangular distribution

ID	Task Name	Duration	Start	Finish	Optimistic Rdur	Most Likely Rdur	Pessimistic Rdur
1	**Single-Path Project**	**220 d**	**6/1**	**1/6**	**155 d**	**200 d**	**305 d**
2	Start	0 d	6/1	6/1	0 d	0 d	0 d
3	Design	55 d	6/1	7/25	40 d	50 d	75 d
4	Build	106.67 d	7/26	11/9	80 d	100 d	140 d
5	Test	43.33 d	11/9	12/22	25 d	35 d	70 d
6	Ship	15 d	12/23	1/6	10 d	15 d	20 d
7	Finish	0 d	1/6	1/6	0 d	0 d	0 d

did that by calculating the standard deviation of the PERT-critical path (this capability is absent from the current versions of Microsoft Project®).[4]

The MOM equation for the standard deviation of the path is the square root of the sum of the variances of the activity durations:

$$\text{Standard deviation of the path} = (\text{Variance}_1 + \text{Variance}_2 + \text{Variance}_3 + \ldots + \text{Variance}_n)^{1/2}$$

Where:

$$\text{Variance} = (\text{Standard Deviation})^2$$

The calculations for the one-path project are shown in Table A.2.

The MOM calculation indicates that the expected number of days is 220, in agreement with the Microsoft Project® calculation above. In addition, however, the standard deviation of 17.5 days is calculated using the MOM equations.

Table A.2   Calculation of mean and standard deviation for the one-path schedule assuming triangular distributions

PERT Calculations using Triangular Distribution						
Activity	**Duration**	**Optimistic**	**Most Likely**	**Pessimistic**	**Mean**	**Variance**
Project	200				220	
Start	0					
Design	50	40	50	75	55	54.2
Build	100	80	100	140	106.7	155.6
Test	35	25	35	70	43.3	93.1
Ship	15	10	15	20	15	4.2
Finish	0		Sum of Variances			306.9
			Standard Deviation			17.5

---

4   The Mean is the 'first moment' of a distribution and the standard deviation is the 'second moment.' There are higher moments, but these are usually sufficient for our purposes.

Using the mean and standard deviation, we can calculate the probability of days other than the mean occurring. First, we have to assume a probability distribution for the total project path completion date.[5] Relying on the Central Limit Theorem, we may assume the distribution of total project completion dates or durations is normal (Gaussian). That means that, with these ranges, there is about a 68 percent probability that the duration of this project is between the mean minus the standard deviation (220-17.5= 202.5 days or December 19) and the mean plus one standard deviation (220 + 17.5 = 237.5 or January 23).

The schedule without risk that ends on December 17 is 200 days in duration. Comparing that date to the normal distribution we see that there is a 13 percent probability of meeting that schedule date (See Figure A.2).[6]

The 80th percentile is January 20[7] as seen in Figure A.3.

These results using the Method of Moments for a single path schedule are very similar to those of Monte Carlo simulation, which showed in Figure 6.4 (Chapter 6) that January 17 is 15 percent likely and that the 80th percentile is January 22. The results shown in Figure 6.4 were derived using the schedule simulation program Pertmaster. The results using Risk+ are similar (see Figure A.4).

Comparing the results of PERT and two Monte Carlo simulation packages for the one-path schedule, where PERT should work well, indicates significant agreement between the methods as expected.

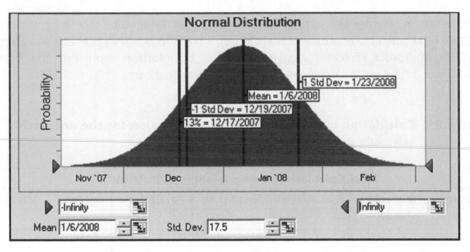

Figure A.2 Normal distribution. Mean of 220 days and standard deviation of 17.5 days

---

5   The Central Limit Theorem says that the mean of the summed durations down a path is the sum of the mean durations of the path's activities, if there are enough activities, and the resulting distribution is normal. What is enough activities? Probably four activities is not enough for strict application of this theorem, but normal is better than other assumptions and it gets more accurate as there are more activities.

6   This chart uses Crystal Ball®, a Monte Carlo simulation engine for Excel from Decisioneering.

7   This result of 13 percent differs slightly from the result from Pertmaster because the distribution of total project completion dates is not strictly normal, as assumed for the PERT calculations.

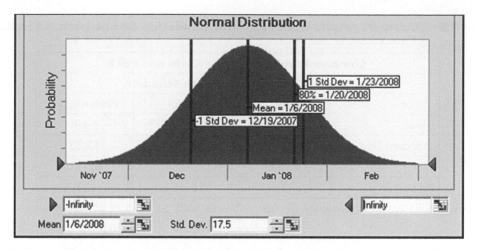

Figure A.3 Checking on the PERT results for the single-path schedule for the 80th percentile; this is January 20

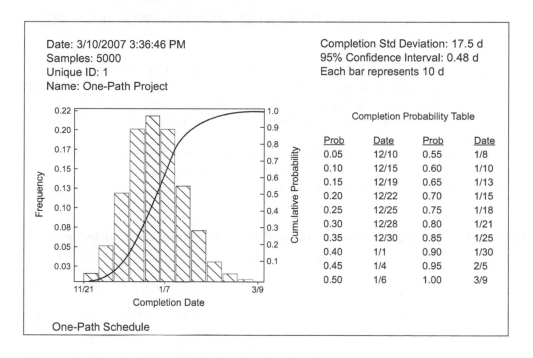

Figure A.4 Results of a Monte Carlo simulation of the simple single-path schedule

Table A.3   Compare results from PERT and simulation on one-path schedule

Compare Results of Monte Carlo and PERT			
One Path Schedule			
	PERT	**Risk+**	**Pertmaster**
Mean	6-Jan	7-Jan	5-Jan
Standard Dev.	17.5d	17.5d	18.3d
P-80	20-Jan	21-Jan	22-Jan
Pr(12/17)	13%	12–15%	15%

Note: PERT results assume a Normal Distribution. Other methods' results make no such assumption.

What Happens With a Real Schedule With Multiple Paths? PERT Underestimates Schedule Risk

ILLUSTRATION OF PERT WHEN IT DOES NOT WORK SO WELL: THE THREE-PATH SCHEDULE

We have shown that PERT will compute reasonable results if the schedule has only one path. However, almost all schedules for real projects have more than one path, and the PERT results are not accurate for those schedules. For this comparison of the PERT and Monte Carlo simulation results we use the three-path schedule introduced in Figure 6.5 (see Chapter 6).

The PERT calculation, still using triangular distributions shows exactly the same results as before with the single-path schedule. This project, too, finishes on average on January 6 for a 220-day duration (see Table A.4).

Table A.4   Three-path schedule using PERT with triangular distributions

ID	Task Name	Duration	Start	Finish	Optimistic Rdur	Most Likely Rdur	Pessimistic Rdur
1	**Three-Path Project**	**220 d**	**6/1**	**1/6**	**155 d**	**200 d**	**305 d**
2	Start	0 d	6/1	6/1	0 d	0 d	0 d
3	**Component 1**	**220 d**	**6/1**	**1/6**	**155 d**	**200 d**	**305 d**
4	Design 1	55 d	6/1	7/25	40 d	50 d	75 d
5	Build 1	106.67 d	7/26	11/9	80 d	100 d	140 d
6	Test 1	43.33 d	11/9	12/22	25 d	35 d	70 d
7	Ship 1	15 d	12/23	1/6	10 d	15 d	20 d
8	**Component 2**	**220 d**	**6/1**	**1/6**	**155 d**	**200 d**	**305 d**
13	**Component 3**	**220 d**	**6/1**	**1/6**	**155 d**	**200 d**	**305 d**
18	Finish	0 d	1/6	1/6	0 d	0 d	0 d

PERT uses the same PERT critical path for its mean and standard deviation calculations, so the results are exactly the same for this three-path schedule as it is for the single-path schedule, 220 days' duration and Finishing on January 6 with a standard deviation of

17.5 days. We have apparently incurred no extra risk by having three risky paths instead of one, if we believe PERT.

Clearly PERT does not show a different schedule between these one-path and three-path schedules. We know intuitively, however, that the three-path schedule must be riskier because there are three paths, each one of which can delay the project (See Figure 6.7 in Chapter 6 for a review of the merge bias). In fact we illustrated that result in the earlier discussion of the merge bias using these simple schedules. The Pertmaster solution to this three-path schedule is shown in Figure 6.6 (Chapter 6). Here the solution using Monte Carlo simulation software Risk+ is shown in Figure A.5.

Comparison of the PERT and Monte Carlo simulation results are shown in Table A.5. They indicate that Monte Carlo has recognized the increased uncertainty in the three-path schedule over the one-path schedule, but PERT has not.

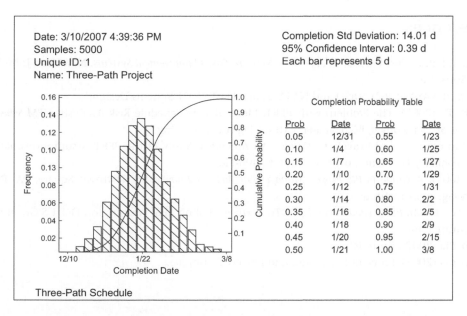

Figure A.5   Monte Carlo simulation results for the three-path schedule

Table A.5   Compare results from PERT and simulation on three-path schedule

Compare Results of Monte Carlo and PERT			
	Three Path Schedule		
	PERT	**Risk+**	**Pertmaster**
Mean	6-Jan	22-Jan	21-Jan
Standard Dev.	17.5d	14.0d	14.3d
P-80	20-Jan	2-Feb	3-Feb
Pr(12/17)	13%	<1%	<1%

Note: PERT assumes Normal Distribution. Other methods' results make no such assumption.

Note that there is close agreement between the Monte Carlo simulation packages but that the PERT results are the same as for the one-path project.[8] PERT has underestimated the risk by concluding that the P-80 is January 20 whereas Monte Carlo correctly concludes February 2-3. Also, PERT has overestimated the probability of finishing on time at 13 percent whereas Monte Carlo correctly estimates schedule success as less than 1 percent.

Unfortunately, PERT underestimates risk at all merge points of the schedule while Monte Carlo accurately handles the merge bias. The extent of underestimation of schedule risk with PERT depends on many considerations, but it is real.

The PERT method has been overtaken by Monte Carlo simulation when used by schedule risk analysis professionals. The general project management profession have adopted the simulation approach and stopped using PERT.[9] Those who still use PERT should explore the more accurate simulation approach.

References

Archibald, R. D. and Villoria, R. L. (1967). *Network-Based Management Systems*. New York, John Wiley & Sons, Inc.

DOD and NASA (1962). 'DOD and NASA Guide: PERT COST Systems Design.'

Hulett, D. (2007). The Problem with PERT: Merge Bias in Schedule Risk Analysis. *CPM Scheduling Turns 50*. Vancouver, BC, Canada, PMI College of Scheduling.

MacCrimmon, K. and Ryavek, C. (1962). An Analytical Study of the PERT Assumptions. *Research Memoranda*, The Rand Corporation. RM-3408-PR.

PMI (2004). *A Guide to the Project Management Body of Knowledge*. Newtown Square, PA, Project Management Institute.

USPOTC (1962). *PERT Fundamentals; POTC Textbook*. Boling AFB, Washington DC, US Government Printing Office.

Wikipedia (2007). Kurtosis.

Wikipedia (2007). Program Evaluation and Review Technique.

---

8   Interestingly, the standard deviation of the distribution for the three-path project from simulation is smaller than for the one-path project. That is because there is very little chance of early dates. With three paths, if one path is short one or both of the other two paths may not be short and the merge-point logic says the schedule is not short. Looked at another way, the cumulative distribution or S-curve for the three-path project stands up straighter than for the one-path project.

9   Again, this is not an indictment of the use of three-point estimating to derive more accurate single-point or deterministic estimates. That method, however, requires collecting the data needed for Monte Carlo simulation, so once the data have been collected, the profession should use simulation methods that work rather than PERT that does not work.

Index

If you have found this book useful you may be interested in other titles from Gower

Benefit Realisation Management
Gerald Bradley
978-0-566-08687-8

Failsafe IS Project Delivery
Andrew Holmes
978-0-566-08255-9

Gower Handbook of Programme Management
Geoff Reiss, Malcolm Anthony, John Chapman, Geof Leigh,
Adrian Pyne and Paul Rayner
978-0-566-08603-8

Gower Handbook of Project Management
Fourth Edition
Edited by Rodney J. Turner
978-0-566-08806-3

The Relationship Manager
Tony Davis and Richard Pharro
978-0-566-08463-8

Tools for Complex Projects
Kaye Remington and Julien Pollack
978-0-566-08741-7

GOWER